D0426273

AMERICAN AND BRITISH TECHNOLOGY IN THE NINETEENTH CENTURY

AMERICAN AND BRITISH TECHNOLOGY IN THE NINETEENTH CENTURY

THE SEARCH FOR LABOUR-SAVING INVENTIONS

BY

H. J. HABAKKUK

Chichele Professor of Economic History and Fellow of All Souls College in the University of Oxford

CAMBRIDGE
AT THE UNIVERSITY PRESS
1967

PUBLISHED BY
THE SYNDICS OF THE CAMBRIDGE UNIVERSITY PRESS
Bentley House, 200 Euston Road, London, N.W.1
American Branch: 32 East 57th Street, New York, N.Y. 10022

First published 1962
Reprinted 1967
First paperback edition 1967

First printed in Great Britain by Adlard & Son Ltd, Bartholomew Press, Dorking
Reprinted, by Lithography, in Great Britain by
Hazell Watson & Viney Ltd,
Aylesbury, Bucks

To my Mother
and to the Memory of my Father

CONTENTS

PREFACE

This essay is a foray into the debatable borderland between history, technology and economics. On the history of technical processes there now exist several works, pre-eminent among them the great five-volumed *History of Technology*. In recent years there have also been considerable advances in the economic theory dealing with the choice of techniques. But few historians of technology have shown much interest in the models of the economists; and the theorists have so far concentrated on analysis or on problems of contemporary technology. A marriage of history and theory could, I believe, be fruitful, and the present work is an attempt to re-examine some of the more familiar nineteenth-century developments in technology in the light of such limited parts of the relevant theory as I feel I understand. It originated in lectures given at Columbia University in the autumn of 1958, and though the first version of the argument has been greatly expanded under the stimulus of criticism and suggestion, I have not attempted to disguise its speculative character. Nor have I taken account of a number of important books, on both the theory and history of the subject, which have appeared since my manuscript was completed.

Anyone who sets up as a middleman is likely to provoke the traditional mistrust of brokers and bodgers, and as I am neither an expert in American economic history nor an economist I expect I have incurred the penalties of those who venture too far beyond their own field. But if I have misread the facts and made a wrong choice or unskilled use of the economic tools, it is in spite of generous help. I am particularly grateful to Kenneth Berrill, Ian Byatt, Conrad Blyth, Jim Potter, Brian Tew, Donald Whitehead and John Wright who read parts of earlier versions of the essay, contributed many ideas and started several trains of thought. I have also picked the brains of N. F. Laing, Guido di Tella and Goran Ohlin, and I learned a good deal from discussion with members of my graduate class in Harvard in 1954–5. My hosts at Columbia, especially Professor Carter Goodrich, made many helpful comments. I also wish to thank my wife for considerable assistance. Confusions and errors are entirely my own.

H. J. HABAKKUK

OXFORD

I

INTRODUCTION

Why did some countries invent and adopt mechanical methods more rapidly than others? Three main types of influence can be distinguished.

(1) First there were sociological influences: the value placed on invention, the amount of inventiveness and natural ingenuity in the population, the extent to which inventive faculties were diverted to industrial purposes by, for example, the educational system; the character of entrepreneurs and their willingness, in response to a given economic inducement, to adopt new methods.

(2) The second influence, of an economic kind, was the volume of capital accumulation. When a large amount of new capacity was created, entrepreneurs had more opportunities to install new techniques and to try out new ideas. The most important part of many improved methods was not the new idea but the accumulation of minor modifications made once the idea had been put into practice. The important consideration for this part of the argument is the absolute amount of new investment rather than the rate of increase, though, other things being equal, the faster the rate the better.

(3) The third influence also had to do with capital accumulation, not because it afforded opportunities for trying out new methods, but because, by running up against bottle-necks in the supply of labour, natural resources or finance, it both induced the early adoption of spontaneous inventions, and stimulated the discovery of new methods. Here what matters is the rate of capital accumulation compared with the rate of growth of co-operating factors. When capital was accumulating more rapidly, shortages emerged in the sense that capitalists had to offer more of their output in order to obtain additional supplies of other factors. The first effect of these emerging scarcities was to reduce the rate of profit and of accumulation in so far as this depended on the rate of profit. But, it is often argued, these restraints proved in some circumstances to be an important stimulus to technical progress, sufficient conceivably to more than offset the unfavourable effects of the fall in the rate of accumulation. The argument has two components:

(*a*) The fall in the rate of profit provided an incentive to look for new methods in *any* direction.

(*b*) But some factors might become scarce sooner than others and so bias the search for new methods in a *particular* direction. Depending on the factor-endowment of the economy, the shortage might show itself first in land, in labour (or in skilled as opposed to unskilled labour) or in raw materials (or in a particular raw material). One type of restraint might prove more fruitful as a stimulus than another.

On the assumption that the volume of investment depended on the rate of profit, and that there was no autonomous technical progress which shifted the technical spectrum, the considerations under (2) worked in an opposite direction from those under (3). The larger the volume of investment the better; on the other hand, scarcities which, in the short run, tended to impede investment might in the longer run have more than countervailing effects on technical progress. There is no *a priori* way of knowing which considerations have been dominant. In the long run do 'restraints', on balance, hold up or accelerate technical development? Are some types of restraints more likely than others to stimulate a technically fruitful response?

This is part of a larger argument, which assumes many forms. Which is more favourable to improvement—expansion or competition? Are entrepreneurs more likely to adopt new methods when they are under pressure, either because factors are scarce or because markets are tight, or when factor-supplies and market prospects are favourable to expansion of capacity? The literature of economic history abounds with conflicting judgements on this problem. Thus abundant labour and abundant finance are often said to have been prerequisites of economic development, but in accounting for the rapidity of German and American growth, historians are apt to assign some of the responsibility, in the case of Germany, to the industrial banks which developed because of a shortage of finance, and in the case of the U.S.A., to mechanisation stimulated by the shortage of labour. These two countries, it is said, were able not merely to produce substitutes for cheap finance and cheap labour but were positively stimulated by these factor-scarcities. Then again historians generally write as if rapidly-expanding markets were favourable to the adoption of new methods, but they are not above ascribing the development of new textile products in seventeenth-century England to the sluggishness of the North-European market for English broadcloth, or the absence of technical progress in the American textile-machine industry between 1840 and 1870 to the softening effect of abundant demand.

There is no reason why abundance of a factor should not have been favourable to technical development in one set of circumstances and scarcity of the same factor favourable in another. The influences which are relevant to development combine in many different ways, and each has a different effect according to the combination in which it appears. Arsenic cures in small doses and kills in large. But this does not dispense with the need to decide which doses are homoeopathic and which are lethal. It is clearly unsatisfactory to say that cheap-labour countries grew because their labour was cheap, and dear-labour countries because their labour was expensive.

This essay is an attempt to discuss the contrasts in technology between the U.S.A. and England in the nineteenth century in the light of these various considerations. Chapters II–V are concerned mainly with the first part of the century, and with the argument that scarcity of labour was responsible for the character of American technology. Chapter II sets the problem. Chapter III considers the terms on which the labour-scarcity explanation can be made internally consistent. Chapter IV is an attempt to speculate, on the basis of the previous analysis, about economic development in the U.S.A. and chapter V does the same for England. Chapter VI discusses the technology of the two countries in the later decades of the nineteenth century.

II

LABOUR-SAVING METHODS IN AMERICAN INDUSTRY: THE PROBLEM

> In Europe work is often wanting for the hands; here (that is, in the United States) hands are wanting for the work. MICHAEL CHEVALIER.[1]

> Where land is very cheap and all men are free, where everyone who so pleases can easily obtain a piece of land for himself, not only is labour very dear, as respects the labourer's share of the produce, but the difficulty is to obtain combined labour at any price. E. G. WAKEFIELD.[2]

There is a substantial body of comment, by English visitors to America in the first half of the nineteenth century, which suggests that, in a number of industries, American equipment was, in some sense, superior to the English even at this period. As early as 1835 Cobden had noted, in the machine shop of a woollen mill at Lowell, 'a number of machines and contrivances for abridging labour greater than at Sharp and Roberts'. He thought agricultural implements in New England exhibited 'remarkable evidences of ingenuity . . . for aiding and abridging human as well as brute labour', and gave several other instances.[3] And the two groups of English technicians who visited America in the 1850's reported that the Americans produced by more highly mechanised and more standardised methods a wide range of products including doors, furniture and other woodwork; boots and shoes; ploughs and mowing-machines, wood screws, files and nails; biscuits; locks, clocks, small arms, nuts and bolts.[4]

[1] Michael Chevalier, *Society, Manners and Politics in the United States* (Boston, 1839), pp. 143–4.

[2] E. G. Wakefield, *England and America* (London, 1833), p. 247.

[3] *The American Diaries of Richard Cobden*, ed. E. Hoon Cawley (Princeton, 1952), pp. 117–18.

[4] *New York Industrial Exhibition; Special Reports of Mr. George Wallis and Mr. Joseph Whitworth*, P.P. 1854, XXXVI. *Report of the Committee on the Machinery of the*

In many and [illegible] probably in most fields of technology the English were still far ahead of the Americans at this date, but [illegible] degree of American mechanisation in a number of [illegible] and the greater stress on standardisation are unexpected enough to call for an explanation; for in 1850 American factory industry was very recent and still a small part of the economy. The development of the American economy after 1870 is not surprising, given the natural resources then available and the nature of the American market. But it is surely surprising that in the 1850's the Small Arms Factory at Enfield should have had to re-equip itself with machine tools made by two American firms at Windsor, in the backwoods of Vermont.[1] Why should mechanisation, standardisation and mass-production have appeared *before* 1850 and to an extent which surprised reasonably dispassionate English observers?

These contemporary comments are usually couched in impressionistic terms, and there is very little data about capital–output ratios, period of production or labour productivity which would allow us to quantify their impressions and to obtain a reasonably unambiguous measure of what it is, in economic terms, that needs to be explained. But that something needs to be explained seems clear.

There were, of course, particular influences in particular cases; for example, the early mechanisation of the U.S. arms industry might be attributed to the particular ability of Whitney and North who developed the system of interchangeable parts; and in looking for pervasive and persistent influences one is, no doubt, in danger of underestimating the force of such local and particular influences, which may in the aggregate have been sufficient to turn the course of development in one direction. But when so many industries are in question it is more reasonable to start by supposing that the explanation must be a fairly general one.

United States, P.P. 1854–5, L, pp. 539–633; see also D. L. Burn, 'The Genesis of American Engineering Competition, 1850–1870', *Economic History* (Supplement to Economic Journal), II (1933). For the extent of mechanisation in the U.S.A. in the 1860's see Charlotte Erickson, *American Industry and the European Immigrant 1860–1885* (Cambridge, Mass. 1957), p. 123. E. W. Watkin, a leading British railway engineer, reported on a visit to the U.S.A. in 1851 that the American railway workshops had 'tools equal or superior to ours in all practical respects'. (*A Trip to the United States and Canada*, 1851, p. 125.)

[1] J. W. Roe, *English and American Tool Builders* (New Haven, 1916), p. 141.

One explanation can be ruled out: in the first half of the nineteenth century the rapidity with which the U.S.A. a[...] labour-saving devices cannot be explained by sup[...] the Americans were investing on a larger scale and therefore had more ample opportunities of trying out and improving new techniques than the English. The American rate of industrial investment in the sections which later became the industrial states may have been higher, but it is obvious that in the industries subject to technical progress the absolute increment of investment (and it is this that matters in the present context) was greater in England, so that English manufacturers as a whole (though not necessarily individually) had more chances of adopting new methods. Both economies had ample opportunities of installing new equipment. The much more important difference between them lay in the restraints which arose as they attempted to expand output. And the relevant question is how far the rapidity of American mechanisation was due to the stimulating effect of bottle-necks, and in particular to a scarcity of labour.

This is often the reason put forward by modern historians. Thus Mr Pelling, for example, observes that it was scarcity of labour 'which laid the foundation for the future continuous progress of American industry, by obliging manufacturers to take every opportunity of installing new types of labour-saving machinery'.[1] This is an echo of what was said at the time. Scarcity of labour was the explanation offered by British observers in the middle of the last century. In 1851, Joseph Whitworth, who had exceptional qualifications for judging, reported after a visit to the States: 'The labouring classes are comparatively few in number, but this is counterbalanced by, and indeed may be regarded as one of the chief causes of, the eagerness with which they call in the aid of machinery in almost every department of industry. Wherever it can be introduced as a substitute for manual labour, it is universally and willingly resorted to. . . . It is this condition of the labour market, and the eager resort to machinery wherever it can be applied to which, under the guidance of superior education and intelligence, the

[1] *British Essays in American History*, ed. H. C. Allen and C. P. Hill (1957), p. 264. See also H. F. Williamson, 'Mass Production, Mass Consumption, and American Industrial Development', in *Contributions to the First International Conference of Economic History* (Stockholm, 1960), pp. 137–43.

remarkable prosperity of the United States is mainly due'.[1] The commission of enquiry into the manufacture of small arms in the U.S.A. concluded that 'in consequence of the scarcity and high price of labour in the United States, and the extreme desire manifested by masters and workmen to adopt all labour-saving devices, from the conviction of such being for their mutual interest, a considerable number of different trades are carried on in the same way as the cotton manufacture of England, viz.: in large factories, with machinery applied to almost every process, the extreme subdivision of labour, and all reduced to an almost perfect system of manufacture'.[2] Similar comments can be found throughout the century. For example, reflecting on the handicraft watch-making industry of Switzerland and the Black Forest, Schulze-Gaevernitz wrote, 'Watch-making, as a home industry, after the model of Switzerland and the Black Forest would be impossible in America, because of the high rate of wages. The factory-system, therefore, became a necessity there'.[3]

But though such expert contemporary comment is to be taken very seriously and the argument seems highly plausible on general grounds, neither the logic of the explanation nor its conformity with the facts are as obvious as at first sight they appear. The only attempt to explain precisely how labour-scarcity was related to mechanisation and technical progress is

[1] P.P. 1854, XXXVI, p. 145. 'Machinery and processes to effect manufactures, so as to leave the manual industry of a nation for other employments, are of a degree of importance to the United States proportioned to the smallness of the average population on a square mile of our extensive territory.' (Tenche Cox, *Digest of Manufactures*, American State Papers, Finance, vol. II, Report No. 407, 5 January 1814, p. 687.)

[2] P.P. 1854–5, L, p. 578.

[3] G. von Schulze-Gaevernitz, *Social Peace*, trans. C. M. Wicksteed, ed. Graham Wallas (1893), p. 125. Comments to similar effect were common throughout the century. 'Very little handwork is done; in the language of the country "it don't pay". A man costs fifteen to twenty dollars a week, but a "man-power" in a steam engine don't cost as many shillings, hence the steam is the cheapest . . . it don't strike, is always at its post, and is easier controlled.' (J. Richards, *A Treatise . . . on Wood-Working Machines* [London, 1872], p. 38.) 'A manufacturer considering the purchase of a machine which will cost 10,000 dollars and replace four labourers but which must pay for itself in ten years, will not hesitate to make the purchase in a country where wages are 500 dollars per annum; here the machine will effect a saving of 1,000 dollars *per annum*. A manufacturer in a country where wages are 200 dollars cannot use the machine, however, because it would cause an annual loss of 200 dollars.' E. Levasseur, *The American Workman*, ed. Th. Marburg (London, 1900), p. 73.

that by Erwin Rothbarth.[1] To attract labour the industrial wage had to be sufficiently high to present an effective alternative to the independent cultivation of land; and such a wage could only be paid if the American industrialist raised the productivity of labour by installing labour-saving machinery. But there are difficulties about this explanation.

If it paid American entrepreneurs to replace expensive American labour by machines made by expensive American labour, why did it not pay English entrepreneurs to replace the cheaper English labour by machines made with that cheaper labour? If this difficulty is resolved, and it is conceded that, because of the high productivity of labour in American agriculture, American manufacturers had a greater incentive to replace labour with machines than did the English, a further objection remains. The high returns in American agriculture imposed a limitation on industrial investment in the U.S.A.; one might expect such industry as existed to make productive use of its labour, but for the same reason one would expect a concentration on land-intensive activities, and the industrial sector to be small and subject to restraints on its growth which were absent in England. This was the expectation of many contemporaries. It was precisely to the existence of uncultivated land still to be had on easy terms that Adam Smith ascribed the fact that no manufactures for distant sale had ever been established in any North American town. 'When an artificer has acquired a little more stock than is necessary for carrying on his own business in supplying the neighbouring county, he does not, in North America, attempt to establish with it a manufacture for more distant sale, but employs it in the improvement and purchase of uncultivated land'.[2] And in the 1820's and '30's the free traders were 'accustomed to point to the higher wages of labour in the U.S. as an insuperable obstacle to the successful establishment of manufactures'.[3] And

[1] E. Rothbarth, 'Causes of the Superior Efficiency of U.S.A. Industry as compared with British Industry', *Economic Journal*, LVI (1946), pp. 383–90.

[2] Adam Smith, *Wealth of Nations*, ed. E. Cannan (1904), I, 358.

[3] F. W. Taussig, *The Tariff History of the United States* (New York, 1914), p. 65. This was a common opinion quite apart from the controversy over the tariff. 'The most prominent of those causes (which impede the introduction and retard the progress of manufactures in the United States) are the abundance of land compared with the population, the high price of labor and the want of a sufficient capital.' (A. Gallatin, *Report on Manufactures:* American State Papers, Finance, vol. II, Report No. 325, 17 April, 1810, p. 430.) 'The outlet for agricultural pro-

indeed there is some reason to suppose that in the eighteenth century the high returns in agriculture *did* have precisely this effect. This was, at bottom, the rationale—as distinct from the emotional force—of Jefferson's dream of a rural republic.

If the argument is extended and it is urged that factor-scarcity may have stimulated the discovery of new methods which more than offset the scarcity, further difficulties arise. Why should scarcity of labour have stimulated a search for methods which saved *labour* as opposed to some other factor? Shortage of a particular factor, it can be argued, did not bias the search for new methods towards those which specifically saved the new factor; all it did was to stimulate the search for any method of reducing costs, and whether this was sought along lines which saved labour or natural resources or capital depended upon technical possibilities, and not upon which factor happened to be scarcest.[1]

But even if this point is waived, and we assume that the scarcity of a particular factor *did* dispose manufacturers to search first for methods which saved that factor, why should a search for labour-saving methods have been more fruitful in technical progress than the search for methods which saved capital or natural resources? Why should a scarcity of labour

duction is so great, and, from the quantity of unoccupied land, the pursuits of agriculture are so easily followed, that there is not much inducement to carry on manufactures west of the mountains except for articles of the first necessity. . . . Agriculture is more profitable and requires less capital.' (*Report of the S.C. on the Export of Machinery*, P.P. 1841, VII, Qu. 2965–6.) 'So long as vast tracts of country of great agricultural capacity remain unoccupied, and while the population continues scarce and labour dear, it cannot be expected that mining and manufacturing enterprises can be carried on to the extents which the wants of the country require.' (*Sir Charles Lyell's Special Report*, P.P. 1854, XXXVI, p. 157.) For a similar view see J. R. McCulloch, *A Statistical Account of the British Empire* (London, 1839), II, p. 13.

[1] Mr G. F. Bloom ('A Note on Professor Hick's Theory of Invention', *American Economic Review*, 1946) argues that, irrespective of the relative price of labour and capital, most invention is naturally devoted to saving labour, because labour is so large an item in total costs and because the desire to make the job easier is the fundamental motive of invention. Inventors may sometimes have supposed that, because a particular factor was the largest cost of a process, the best chances of cutting costs lay in saving that factor. But as a general assumption this would have been irrational. There are, moreover, instances of invention which were specifically designed to save natural resources. Nor is it at all certain that labour-saving inventions predominate in all periods. Professor Solow concludes that in America, 1909–49, technical change was on the average neutral as between capital and labour. (R. M. Solow, 'Technical Change and the Aggregate Production Function', *Review of Economics and Statistics*, XXXIX [August 1957], p. 320.)

in relation to land and capital in the U.S.A. have been a greater stimulus than the scarcity of land and capital in relation to labour in England? Why should factor-endowment have asymmetrical effects?

Finally, even if we attach considerable importance to the influence of factor-endowment on the invention and adoption of new methods, it cannot be assumed that it was differences on this score which were responsible for the salient differences in English and American technological development. Granted that, at the same rate of investment in relation to total factor supplies, labour was a scarcer factor in the U.S.A. than in England, and also that a scarcity of labour had a more generally stimulating effect than an equivalent scarcity of any other factor, the possibility nevertheless remains that the crucial difference was not here, but in the rate of investment. If, for quite independent reasons, industrial investment, in relation to factor-supplies, was increasing more rapidly in the U.S.A. than in England, scarcities of other factors as well as labour were likely to arise in the U.S.A., and her industrial development would therefore have been more sensitive to factor-endowment.

III

THE ECONOMIC EFFECTS OF LABOUR-SCARCITY

In this chapter we examine the logical implications of the labour-scarcity argument under the following heads:

(*a*) Labour-Scarcity and the Choice of Technique.
(*b*) Labour-Scarcity and the Rate of Investment.
(*c*) Market Imperfections.
(*d*) Labour-Scarcity and the Trade Cycle.
(*e*) The Expansion of the Market.

LABOUR-SCARCITY AND CHOICE OF TECHNIQUE

The supply of labour in the U.S.A. and England

Industry started to develop in the United States at a time when industrial money wages were substantially higher than they were in England, according to some estimates for the early nineteenth century perhaps a third to a half higher.[1] This was fundamentally because the remuneration of American industrial labour was measured by the rewards and advantages of independent agriculture.[2] 'Where land is in such plenty',

[1] In the colonial period American 'workmen commanded wages from 30 to 100 per cent higher than the wages of contemporary English labouring men.' (*The Growth of the American Economy*, ed. H. F. Williamson, pp. 49–50, 53, 101, 137.) According to Nassau Senior, writing in 1829, labourers' wages in North America were 'twenty-five per cent higher than they are in England, while the labour requisite to obtain necessaries is not much more than half as great in the former country as in the latter'. (*On the Cost of Obtaining Money* (London, 1830), pp. 2, 11, London School of Economic Reprints No. 5.) 'That the general rate of wages is higher in the United States than in Britain is admitted, particularly the wages of females employed in the Factories.' James Montgomery, *A Practical Detail of the Cotton Manufacture of the United States . . . Compared with that of Great Britain* (Glasgow, 1840), p. 135.

[2] V. S. Clark, *History of Manufactures in the United States* (Washington, 1929), vol. 1, pp. 152–8. The clearest statement of this connection was put by Hamilton in the process of arguing against it. 'The smallness of their population compared with their territory; the constant allurements to emigration from the settled to the unsettled parts of the country; the facility with which the less independent condition of an artisan can be exchanged for the more independent condition of a farmer; these and similar causes, conspire to produce, and, for a length of time,

wrote an observer in the 1760's, 'men very soon become farmers, however low they set out in life. Where this is the case, it must at once be evident that the price of labour must be very dear; nothing but a high price will induce men to labour at all, and at the same time it presently puts a conclusion to it by enabling them to take a piece of waste land. By day-labourers, which are not common in the colonies, one shilling will do as much in England as half-a-crown in New England.'[1] Land was abundant and, except possibly in Virginia before the abolition of entails, it was accessible to purchase by men of small means. In the 1820's it could be purchased from the Federal Government at $1.25 per acre, which was well within the reach of a

must continue to occasion, a scarcity of hands for manufacturing occupation, and dearness of labour generally.' (A. Hamilton, *Report on Manufactures*, p. 123.) 'The price of manual labor, and the compensation of what is purely mechanical, such as the salary of ordinary clerks, is much higher than in Europe. This is a simple result of the comparative state of demand and supply for those objects, arising principally from the superabundance of land compared with the present population.' Gallatin to La Fayette, 12 May 1833, *Writings of Gallatin*, ed. Henry Adams (Philadelphia, 1879), II, p. 471. To argue that, in the first half of the nineteenth century, agriculture set a floor to American industrial wages, it is not necessary to suppose that large numbers of industrial workers moved west. It is sufficient if 'the abundance of western land drew away many thousands of *potential* wage earners (from the hill towns of New England and from the exhausted farms of New York and Pennsylvania) who might otherwise have crowded into the factories'. (C. Goodrich and S. Davidson, 'The Wage-earner in the Westward Movement,' *Political Science Quarterly*, LI [1936], p. 115.) Thus the argument does not depend upon Turner's frontier hypothesis, rigorously interpreted. A large amount of work has been devoted to demolishing this thesis; the work that is most relevant to our argument attempts to show (*a*) that the 'West' lured more Europeans to the Eastern states than Easterners to the West, and (*b*) that there was a net migration from farming families to the towns. (See F. A. Shannon, 'A Post-Mortem on the Labour-Safety-Valve Theory', *Agricultural History* (1945). The demonstration has been made for the period *after* 1860 and therefore does not bear directly on our argument. Moreover, even for the period after 1860, Professor Shannon's conclusions are likely to underestimate the effect of returns in agriculture on wages. He compared the actual farming population in 1900 (28 million) with his estimate (40 million) of the population which would have been produced by the natural increase of the farming population of 1860 (19 million), and he concluded that the excess of 40 millions over 28 millions represented migration out of agriculture. But his estimate of 40 millions was arrived at by multiplying the farming population of 1860 by an 'average rate of breeding' which was not, in fact, the rate of natural increase but *included* increase by immigration. Moreover, in his calculations, he restricted the frontier to the 'farming frontier'; but from the point of view of our argument the 'pull' of the frontier towns, which develop because of the movement of the farming frontier, is also relevant. On this subject I am indebted to William Fredrickson.

[1] *American Husbandry*, ed. H. J. Carman (New York, 1939), pp. 53–4, quoted by D. L. Kemmerer, 'The Changing Pattern of American Economic Development' *Journal of Economic History*, XVI, 1956, p. 576.

skilled labourer, who might at this time earn between $1.25 and $2.0 per day.[1] 'The men earn here (in the cotton-textile factories at Lowell) from 10 to 20 dollars a week', wrote an English observer in 1842, 'and can therefore lay by from 5 to 10 dollars, after providing for every want, so that in two or three years they accumulate enough to go off to the west and buy an estate at 1¼ dollars an acre or set up in some small way of business at home'.[2] In England, by contrast, land was scarce in relation to labour, and the supply of land on the market, particularly of small properties, was restricted by the existence of large estates supported by legal restraints on alienation; the return on the cultivation of land in England up to 1815 was high since new techniques were available, and food prices were rising; but to set up as a tenant-farmer required considerable capital, and, even if an English artisan had been able to acquire the capital, the supply of farms to be let was limited, and absorbed by the demand of the sons of existing tenants. In England, therefore, a man could, generally speaking, enter agriculture only as a labourer commanding low wages.

The abundance and accessibility of land plus the fact that much of it was fertile meant that output per man in American agriculture was high. Moreover, since the cultivator was often also the owner, and his family supplied the labour, the advantage of the high output accrued to the cultivator. His income included: (1) an element of rent, which would in England have been a heavy charge on output and payable to the landowner, (2) agricultural profits, which in England accrued to the tenant-farmer, as well as (3) the wages, which in England went to the agricultural labourer. Furthermore the new land was brought into use in such a way that the returns to settlement on the frontier sometimes included elements of exceptional gain. Many American farmers had heavy debt charges; but there

[1] It was available at this price after 1820 if bought in tracts of 80 acres (the units previously laid down were 640 acres (1785), 320 acres (1800), 160 acres (1804). There were in addition the costs of bringing it under cultivation. 'Buying the land is nothing; the real expense is getting it ready, on account of the cost of labour.' In Michigan in 1831, one of de Tocqueville's informants estimated that to get the land under cultivation cost 5 to 10 dollars per acre. Another settler gave a lower estimate: to establish a new settlement one needed between 150 and 200 dollars, 100 of which was to buy the 80 acres and the rest for the expenses of initial settlement and incidentals. (de Tocqueville, *Journeys to England and Ireland*, ed. J. P. Mayer (1958), pp. 209–10, see also p. 343.

[2] J. S. Buckingham, *The Eastern and Western States of America* (1842), I, pp. 300–1.

were no tithes, and taxes were low. Finally since the total earnings of the family in American agriculture tended to be divided among its members, there was less disparity between average and marginal earnings. In order to attract labour, therefore, industry had to assure the workers in industry a real wage comparable to average earnings in agriculture. English industry, by contrast, could acquire labour from agriculture at a wage equal to the very low product of the marginal agricultural labourer plus an addition to cover the costs of transport and of overcoming inertia. Thus, while in England industrial wages equalled the marginal product, in the U.S.A. the reward of the marginal labourer in industry was above his product, unless the manufacturer took steps to increase the product.

Moreover the course of agricultural technology in the early decades of the nineteenth century may well have accentuated the disparity between the terms on which labour was available to industry in the U.S.A. and England. In America improvements in agriculture took the form primarily of increasing output per head and the increase initially was probably more rapid than in industry; in England on the other hand, agricultural improvement was devoted primarily to increasing yields per acre and, even where there was an increase in output per head, the abundance of labour made it difficult for the labourer to enjoy the increase. In America agricultural improvements raised, and in England prevented, a rise in the terms on which labour was available to industry.

Comparison of industrial wage-rates does not of course measure precisely the difference in labour-cost, in terms of output, in the two countries. For, on the one side, the hours of work in America were generally longer,[1] and conceivably effort was more concentrated. It may also be, as some contemporaries said, that better nutrition and more spacious working and living conditions, made the American a more efficient worker. On the other hand it might be supposed from the alternatives open to labour in America that the workers recruited into American industry were inferior, in relation to agricultural labour, when compared with the English; it is possible that is, that the pull of agriculture showed itself in the quality as well as the price of

[1] 'The number of hours constituting a day's work especially in factory labour, is much greater; and. . . this excess in the hours of work obtains generally in every industrial occupation.' (P.P. 1854, XXXVI, p. 14.)

industrial labour in America. Then again, American employers had to incur higher costs than the English on housing and working-facilities of a kind which made the work more agreeable but added little, if anything, to labour productivity. Probably also the rate of turnover of workers was higher in America, and therefore the likelihood smaller that they would acquire industrial discipline. Finally, England at the start was technically superior and this must have reflected itself in higher labour productivity. Since we cannot measure these conflicting influences, it is impossible to be precise about the differences in labour-costs in the two countries, but there is no reason to doubt the opinion of contemporaries that American industrial labour was substantially dearer than the English.

But American industrial labour was not only dearer than the English; its supply was less elastic.[1] It was more difficult to obtain additional labour for the industrial sector as a whole. This was partly owing to the abundance of land and the diffi-

[1] M. D. Morris ['The Recruitment of an industrial Labor Force in India, with British and American Comparisons', *Comparative Studies in Society and History*, II (1960), pp. 305–28] argues that in neither the U.S.A. nor Britain did the recruitment of labour significantly inhibit industrial expansion in the cotton-textile industry, but concedes that in the U.S.A. 'during the periods when the industry was expanding rapidly the new labor force did not always automatically flow in as rapidly as was necessary to fill the mills with labor. In this sense, the situation was different (from the) typical British case' (p. 319). It is impossible to say how great a disparity in this respect existed between the two countries, especially since there were considerable regional variations within the U.S.A. itself. But that there was some disparity seems a fair deduction from contemporary comment. 'Employment (in America) . . . is always plentiful: labour, especially skilled labour, ever in demand.' 'Full employment can always be obtained by competent workmen.' (P.P. 1854, XXXVI, pp. 14, 46.) The American manufacturer 'must often carry on his business at little or no profit, perhaps at a considerable loss, in order to keep together the agents by which, when the demand comes, he can alone supply it. Hence the anxiety to settle down the operatives around the mills; to render their condition and social position such as shall absolutely attach them to their employers and the locality. The latter, perhaps, being the most difficult, from the migratory tendencies of a people so restless, and always so alive to any new contingency which promises to better their condition, however distant the field of operation may be'. (P.P. 1854, XXXVI, p. 42). It was particularly difficult to get together enough skilled operatives after a firm had reduced its operations, because they sought and obtained employment in other districts. '. . . in Boston a constant deficiency of labour had seriously hampered the growth of industry until the forties. . . . Few (industries) grew significantly (1835–1845), and many actually declined. The prospective manufacturer desiring a site for a new establishment, or the capitalist with an "abundance of money seeking an outlet" found little encouragement. And even those already established who wished to expand were inhibited by the apparently inflexible labour supply'. (O. Handlin, *Boston's Immigrants* [2nd ed. Cambridge, Mass. 1959], p. 74.)

culties of internal transport, which required technical solutions and heavy capital outlays before they could be overcome. It was also partly owing to America's geographical remoteness from the areas of abundant population. The U.S.A. in the early decades of the nineteenth century was not only sparsely populated, but was forced by the high costs of transport (including the costs of overcoming inertia) to depend for labour, down to the 1840's, mainly on her own resources. And though the American rate of natural increase of population was exceptionally high, a large part of it was absorbed in settling new land, and the labour to work additions to industrial capacity tended to be difficult to obtain. The Superintendent of the Springfield Armoury wrote in 1822 'I find the rage for manufacturing cotton prevails to such a degree and there is so great a call for first-rate workmen that I am apprehensive I shall lose some of our most valuable workmen except I am authorised to raise their wages according to circumstances'.[1] By contrast, the industrial areas of England were near to densely-populated agrarian slums in the southern agricultural counties and in Ireland. For an identical rate of increase of capacity, therefore, the American industry would have had more difficulty than the English in obtaining additional labour.

Dearness and inelasticity are logically distinct; the consequences of a high level of wages prevailing over time are not the same at all points as those of a wage-level which rises when demand for labour increases. There have been situations where the floor set to industrial wages by *per capita* productivity in agriculture was low but the supply of labour was inelastic, for example in England for much of the eighteenth century. There have also been situations where the floor set to industrial wages was high but where abundant labour was forthcoming at this wage—in some respects this was the case in the U.S.A. in the second half of the nineteenth century when there was a large amount of immigration. But in the first half of the nineteenth century, with which we are here mainly concerned, there was a contrast on both points, and a more marked contrast than existed either before or since; industrial labour was dear, and its total supply inelastic in the U.S.A. and it was cheap and elastic in England. Since the general level of labour-costs was so

[1] F. J. Deyrup, *Arms Makers of the Connecticut Valley* (Smith College Studies in History, vol. XXXIII; Northampton, Mass. 1948), p. 105.

closely connected with the elasticity of the supply of labour, it is difficult to discuss their effects separately except at the cost of repetition. In most of the following discussion, therefore, they are treated together.

The inducement to mechanise

It seems obvious—it certainly seemed so to contemporaries—that the dearness and inelasticity of American, compared with British labour, gave the American entrepreneur with a given capital a greater inducement than his British counterpart to replace labour by machines. The real problem is to determine how the substitution took place. Where the more mechanised method saved *both* labour and capital per unit of output it would be the preferred technique in both countries. It was where the more mechanised method saved labour but at the expense of an increase in capital that the American had the greater inducement than the English manufacturer to adopt it. [The term capital-intensive will henceforward be used to describe such a method.]

A number of situations can be distinguished in which this inducement to make a more capital-intensive choice of technique would operate.

(*a*) Provided final product-prices did not rise—or, at any rate, did not rise in the same proportion—a given rise in wages in terms of output reduced the rate of profit in the method with a low output per man but low capital-intensity more than in the method which had a higher output per man because it was more capital-intensive, *even if* the price of machines rose in the same proportion as wages. If the price of machines did *not* rise in the same proportion as wages, the relative advantage of the capital-intensive method would, of course, be even more pronounced. Given the assumption about product-prices, this is simply a matter of arithmetic.[1] The effect of a rise in wages in a situation broadly analogous to the one just described was first analysed by Ricardo, and in his model product-prices remained constant (as well incidentally as machine-prices) because, in his view, the general level of prices was determined by the gold

[1] J. Robinson, *The Accumulation of Capital* (1958), p. 107. D. H. Robertson, *Lectures on Economic Principles* (1958), pp. 110–13. In this section I have drawn heavily on Mrs Robinson's book.

supply.[1] If one rejects Ricardo's view of the way in which the general price level is determined, it is not easy to see why, in a closed system and the long run, the price of products, like the price of machines, should not rise in the same proportion as wages, in which case there would be no inducement to shift towards more capital-intensive methods. But, in real life, it would take time before the rise in wages worked itself out, and the price of final products and also of machines would lag behind. For some time, most final products would be produced by the existing stock of machines, made with labour before the rise in wages, and the same would be true of new machines. In a world where price was equal to marginal cost this would mean only that manufacturers who produced with the existing stock of machines would enjoy a period of quasi-rents. But in practice it is unlikely that prices would rise to this extent. Moreover we are not considering a closed system, and it is reasonable to assume, for the American economy in the first half of the nineteenth century, that product-prices did not uniformly rise in proportion to the cost of labour in the case of those products in which there was an international price to which American prices tended over a period to accommodate themselves; that is, when imports prevented the American producer passing on the rise in his labour-costs to the consumer.

Moreover, a similar bias towards the more capital-intensive methods would be exerted if, when product-prices fell—because of a deflation of demand—money wage rates did not fall proportionately; and it is probable that the great alternatives to industrial employment in the U.S.A., while they compelled the manufacturer to pay much higher wages for additional labour, also made it difficult, in the short run, to reduce wage rates when product-prices fell, for one consequence of labour-shortage in America was that labour was often hired by yearly contract.[2]

[1] D. Ricardo, *Works and Correspondence*, ed. P. Sraffa (Cambridge, 1951), I, ch. XXXI p. 395. 'The consequence of a rise in food will be a rise in wages and every rise of wages will have a tendency to determine the saved capital in a greater proportion than before to the employment of machinery. Machinery and labour are in constant competition, and the former can frequently not be employed till labour rises.'

[2] 'The greater part of the (cotton textile) Factory workers being connected with farming, whenever wages become reduced so low as to cease to operate as an inducement to prefer Factory labour above any other to which they can turn their attention, then a great many Factories will have to shut up. During a stagnation

(*b*) The argument in the preceding paragraphs has assumed that the construction-time of machines was so short that it can be ignored, and we have not considered the effect on the relative cost of different techniques of the rate of interest in the calculation of capital-costs. We shall now consider what happens when this assumption is dropped. The cost of capital can be considered as consisting of two parts, the principal, that is the price of the plant, and the rate of interest which the manufacturer applied to this principal. If the principal is thought of as consisting solely of labour, it would rise in the same proportion as the rise in the price of labour. But, on Ricardo's assumptions, the total interest bill would not rise in the same proportion, because the rate of interest applied to the capital would now be lower. If manufacturers, in calculating the cost of capital goods, employed a notional rate of interest equal to the rate of profit, this would in general tend to increase the relative attractions of the more capital-intensive technique, because capital would be cheapened relatively to labour.

It is, however, important to notice that there are certain circumstances in which, on our present assumptions, a rise in wages might have the reverse effect, that is, might induce the adoption of a technique which was less capital-intensive and less productive per unit of labour, though still the most profitable in the circumstances.[1] And since this possibility is of more than theoretical interest it needs to be explored. Capital-cost consists of the outlays on new plant plus the interest on these outlays during the construction-period of the plant minus the interest on the amortisation funds out of the earnings of the plant during its life-time. At the higher level of wages, the rate of profit and therefore, on the present assumptions, the rate of interest employed in calculating capital-costs, would be lower than before. There are circumstances in which, at this lower rate of interest, the less capital-intensive technique will be preferable; though its output per man is less, its capital-costs are now lower in a greater degree. When the capital-intensive

in trade, it is common for the manufacturers here (i.e. in the U.S.A.) to stop a part, or the whole of their Factories, and then the workers retire to their farms; such was the case in 1837, when a vast number of Factories were entirely shut up.' (J. Montgomery, *The Cotton Manufacture of the United States contrasted and compared with that of Great Britain* (Glasgow, 1840), p. 137.)

[1] See Joan Robinson, *op. cit.* pp. 109–10, and C. A. Blyth, 'Towards a More General Theory of Capital', *Economica*, XXVII, May 1960, pp. 126–8.

technique has a much longer construction-period, in relation to its life-time, than the labour-intensive technique, that is when it locks up capital for a long period before any output is forthcoming, the entrepreneur at the higher level of wages may find it more profitable to employ the technique which has the shorter construction-period in relation to its life-time, that is, which starts to yield its output sooner, even though the output per unit of current labour is lower.

The circumstances in which this may occur have been analysed. The first condition is that the choice lies between techniques which vary considerably in the ratio between their construction-time and their life-time, the construction-period being longer, in relation to life-time, in the capital-intensive than in the labour-intensive technique. The second condition is that the relation between construction-time and life-time is fixed within narrow limits by technical considerations, so that the techniques available do not form a spectrum and the entrepreneur has to choose between radically different ways of using his resources. But how frequently are these circumstances likely to have occurred in practice? The theoretical case has been illustrated by a comparison between the production of meat by the fattening of cattle and by the grazing of sheep, the first standing for a process which yields its product only after a long period of expensive preparation, but then yields a very large product for the labour employed, the second a process which yields a small output per man, but yields it sooner and more continuously.[1] At a high level of wages, the balance of advantage may turn against the more capital-intensive labour-productive method which locks up its capital for a longer period. Agriculture might provide other instances, for shifts from one crop to another, or from arable to pasture, necessitate radical changes in organisation. There may sometimes be analogous choices in mining or in the production of primary products where the plantation is an alternative to peasant agriculture.

A rise in wages *may*, then, provide an incentive to adopt techniques which are less capital-intensive. But this effect is a possibility only where a longer construction-time is linked, by technical necessity, with a shorter life-time. And it is difficult to think of examples in manufacturing proper. In manufacturing,

[1] Blyth, *op. cit.*, pp. 126–8.

there is more likely to be a continuous range of known techniques, such that the entrepreneur can move smoothly along the range, making small adaptations all the time; in so far as there are differences in this respect, it is generally the more capital-intensive and labour productive of the known methods which are the most durable, that is, have the longer life-time in relation to their construction-period. While therefore it is not a universal rule that a rise in wages will change the ranking of techniques according to profitability in favour of those which are both more productive per unit of labour and more capital-intensive, it is justifiable to suppose that this is the common case in industry.

(*c*) In paragraphs (*a*) and (*b*) above we have been concerned with variations of the 'Ricardo effect'. But quite apart from this effect, a rise in wages would have raised the cost of capital-intensive techniques less than that of labour-intensive techniques if it was easier to import machines than labour from countries where labour was cheaper and its supply more elastic. Even if wages in America rose in the same degree for all types of labour, the attractions of the more capital-intensive methods would still have become relatively greater to American manufacturers, as a result of the rise in their labour-costs, if they could import capital-intensive equipment from areas where labour-costs had not risen.

(*d*) There may also have been significant differences in the elasticity of supply of different *types* of labour and in the extent to which capital- and labour-intensive techniques used the different types; and where the more capital-intensive techniques made relatively greater use of the type of labour which was most abundant, a general rise in wages would favour their use.

There is little readily available information about the labour requirements of various techniques or about the price of different types of labour, and the following discussion is therefore conjectural. But a plausible case can be made for supposing that in the early nineteenth century in the U.S.A. an increased demand for labour raised the wages of skilled labour *less* than the wages of unskilled labour, and that, in many cases, the capital-intensive technique required, for its construction plus operation, more skilled labour per unit of output than the labour-intensive technique.

Was the differential for skill smaller in America than in England? V. S. Clark, referring apparently to the 1820's, considered that differences in wages between England and America were greater in unskilled than in skilled occupations.[1] On the other hand some incidental comment suggests that it was difficult to train a class of skilled workers in America at least in textiles. 'During the first half of the (nineteenth) century, when the ring-frame was being introduced and when the operatives were native-born, the labour-force in the mills was constantly changing . . . so that no skilled class was developed.'[2] And Whitney described the leading object of the system of interchangeable parts as 'to substitute correct and effective operations of machinery for that skill of the artist which is acquired only by long practice and experience; a species of skill which is not possessed in this country to any considerable extent'.[3]

The distinction between skilled and unskilled labour is, of course, constantly shifting; technical progress creates new categories of employment and calls for continual redefinition of skill. Skilled labour at the beginning of the nineteenth century was very different from skilled labour at the end. At the end of the century there was a whole spectrum of degrees of skill. At the beginning of the century there were three broad categories of labour. First there was the undifferentiated mass of unskilled adult labour. The money wage of such labour was a third or a half higher in America than in England. Secondly there were workers who performed tasks which required dexterity and aptitude but which, granted these qualities, could be performed after a short period of training and experience, for example some of the tasks performed by women in the textile industry. For such labour the American wage was rarely more than 20 per cent higher than the English.[4] Finally there were the craft skills which were so technically complicated that they could be acquired only after a long term of training. Craft operations were so diverse that, more than in other types of labour, it is extremely difficult to make direct comparison, especially as

[1] V. S. Clark, *op. cit.* I, p. 392. For further discussion of this point see below pp. 128–131, 151–6.

[2] M. T. Copeland, *The Cotton Manufacturing Industry of the United States* (Cambridge, Mass. 1912), p. 73.

[3] Quoted in J. W. Roe, *op. cit.* pp. 132–3.

[4] V. S. Clark, *op. cit.* p. 397.

rates varied widely according to season and from place to place, and we have no independent tests of the degree of skill being priced. Only a very detailed analysis of labour capabilities and of the relative values placed upon them in the two countries would establish the differences in this respect between America and England. But the random selection of rates given by Clark suggests that the premium on artisan skills was generally lower in America than in England in the early nineteenth century.

How far differences in the premium on skill between the two countries represent difference in supply and how far differences in demand it is impossible to say, but there are some general reasons why we should expect the supply of skill, in relation to common labour, to have been more abundant in America than in England:

(1) As has been evident since 1939, a general shortage of labour is most acutely felt in the unskilled grades, in a shortage of recruits for heavy tedious work; workers in low-paid activities are more prepared to leave their jobs and seek better ones when labour is scarce, in relation to demand, than when it is abundant. A general shortage of labour raises the labour-costs of instrument-users more than those of instrument-makers. Where there is a persistent surplus of labour, it is those who are without skill, particularly the newcomers to the labour market, who have most difficulty in finding work; it is on the wages of the unskilled that the surplus has most effect.

(2) The pulling or retaining power of American agricultural expansion was felt most on unskilled labour. It was, of course, easier for the skilled worker to accumulate the capital necessary for settlement; but at the opening of the nineteenth century the costs of settlement were probably sufficiently low in relation to industrial earnings not to restrict the possibility to the highest-paid workers, and it was the worker without special industrial skills who stood to make the largest relative gain from agriculture. Furthermore, investment in social overhead capital, particularly transport-systems, made heavy demands on general labour, that is labour not trained for particular operations, and the construction of canals, roads and railways seems to have been more attractive to such labour than the factories. This type of investment was a more rapidly increasing proportion of total investment than in England.

(3) Literacy was more widely diffused in America and popular education developed earlier. 'There are very few really *ignorant* men in America of native growth'[1] wrote Cobbett. Thus a higher

[1] Cobbett, *op. cit.* p. 197.

proportion of the population than in England was capable of being trained to skilled operations.

(4) There was much more international mobility of skill than of general labour, and a high proportion of English migrants to the U.S.A. before the start of mass migration were skilled workers. In the early decades of the century therefore immigration did more to alleviate the shortages of artisan skills than of unskilled labour.

(5) Mechanical abilities of a rudimentary sort were widely spread in the U.S.A. at the opening of the century. 'The manufacturing enterprises which existed in the heart of America's eighteenth-century mercantile-agricultural economy were numerous and diversified . . . varied and dextrous mechanical abilities were all but universal.'[1]

(6) The up-grading of unskilled to skilled labour was less impeded in the U.S.A. because, though skilled workers were in fact organised earlier in the U.S.A., trade union restrictions, conventions, apprenticeship rules, were less effective than in England.

We may reasonably conclude, therefore, that in America an increase in the demand for labour raised the cost of the methods which required a great deal of unskilled labour more than it raised the cost of the methods which required a great deal of skilled labour.

It is not always the case that the capital-intensive methods require more skilled labour per unit of output than the labour-intensive. But in the technology of the early nineteenth century there are likely to have been several cases where it did so. The *manufacture* of power-looms required more skill than the manufacture of hand-looms; and the same was probably true of the 'superior' as compared with the 'simpler' machines of all kinds. In the U.S.A. when demand for labour rose, the labour-costs of the machine-makers rose less than the labour-costs of the machine-users, and the costs of the machines which were expensive in terms of output rose less than the cost of the cheaper machines. At the very start, of course, 'power-looms' may have been more expensive in terms of 'hand-looms' in America than in Britain, because American deficiencies in engineering skill, compared with the British, were more marked in the making of complicated than of simple machines. The point is that, with the increase in industrial capacity, the ratio between the costs of manufacturing the equipment fell

[1] G. S. Gibb, *The Saco-Lowell Shops: Textile Machinery Building in New England 1813–1949* (Cambridge, Mass. 1950), p. 10.

more rapidly in America than in Britain, quite apart from any possibility that the Americans were catching up on the English in engineering skills.

The position of the *operating costs* of capital-intensive and labour-intensive techniques is not so clear. There may have been industries in which the operation of the capital-intensive machine required a *lower* ratio of skilled to unskilled labour, which, to a greater or lesser degree, offset the higher ratio in the costs of its manufacture. But the probability is that, in a significant number of cases, the manufacture plus use of the more capital-intensive techniques required more skilled to unskilled labour than the labour-intensive. Where this was so, the fact that unskilled labour, in relation to skilled labour, was dearer in America than in England gave the American an inducement to make a more capital-intensive choice of technique. There is also the additional point that the type of labour which was relatively dearest performed the simple, unskilled operations which were, from a technical point of view, most easily mechanised.

Thus there were at least four circumstances in which a rise in the cost of American labour provided the American manufacturer with an incentive to adopt the more capital-intensive of known techniques, in order partly to check the rise in wages and partly to compensate for it. As, from experience, it became evident that labour was the scarcity most likely to emerge during a general attempt to expand capacity, we should expect an increasing number of American manufacturers to have this in mind when choosing equipment and to become conditioned to adopting the method which did most to alleviate this particular scarcity. It is not, however, strictly necessary to assume that all, or indeed any, manufacturers consciously reflected in this way on the resource-saving characteristics of different techniques. Investment may have adapted itself to relative factor-scarcities by a process of natural as well as of conscious selection. If some manufacturers, for whatever reason, adopted improvements which were more appropriate to the factor-endowment of the economy, these men fared better than those which made a contrary choice. They competed more successfully in product- and factor-markets and by expanding their operations came in time to constitute a larger share of their industry.[1] Moreover

[1] W. Fellner, *Trends and Cycles in Economic Activity* (New York, 1956), p. 50.

they brought influence to bear not only via the market but by force of example. Their success inspired imitators, and shaped entreprenurial attitudes towards the most likely lines of development.

In England, where the supply of labour to industry as a whole was elastic, there was no reason, so far as labour-supplies were concerned, why accumulation should not proceed by the multiplication of machines of the existing type. Thus even if the general level of labour-costs had not been higher in America, the difficulty of attracting additional labour might have pushed the American entrepreneur over a gap in the range of techniques and induced him to adopt one which was not only more capital-intensive than he had previously employed, but was also more capital-intensive than those currently adopted by his English counterpart. But the fact that American labour was also dearer than the English provided the American entrepreneur with an incentive to adopt more capital-intensive techniques than the English, even had the supply of labour been equally forthcoming.

For this argument it does not seem essential that the cost of finance in industry should have been lower, in relation to labour-costs, in the U.S.A. than in Britain. It is enough if machines were cheaper in America in relation to labour, either because they could be imported or because they were made with the type of labour which was relatively most abundant. But the bias towards capital-intensity would obviously have been greater if in fact finance was cheaper in relation to labour in America.

The cost of finance to a manufacturer who reinvested his profits is more ambiguous than the cost of labour, because we do not know what, if any, imputed rate of interest was used. It follows from the assumptions of the simplified version of events to which we have previously referred that the country with the higher wages in terms of output has the lower rate of profit on capital, and this can be regarded as, in some sense, the relevant rate of interest. But this is not a very helpful guide to what actually happened. For though there are many quotations of profits for particular firms in particular years, it is rarely clear what definition has been employed and it is difficult to derive any general rate of profit.[1] In any case accounting methods in

[1] There is a summary of such data in Clark, *op. cit.* I, pp. 373–8.

early nineteenth-century manufacturing were extremely rudimentary, and it is wildly unlikely that, in comparing alternative techniques, anyone ever applied a notional rate of interest equal to the anticipated rate of profit.

For many of the simple techniques of the period it probably did not matter much if the entrepreneur neglected to impute a rate of interest on his locked-up capital. When the capital per unit of output was substantial, as it may have been in the construction of cotton mills, the need to impute some rate of interest would have been more evident, and probably rule-of-thumb methods were devised which corresponded well enough in general effect with more rigorous accounting principles. It is possible that manufacturers reckoned the cost of their capital by the alternative uses to which they could put their savings. Over a country so large and diverse it is impossible to say much in general terms about the consequences of proceeding in this way. The opportunity-costs of industrial finance in America were clearly higher in the U.S.A. as a whole than in Britain. But the extremely high rates of interest which are sometimes quoted, for example on business paper, reflect a considerable degree of risk and other market imperfections, and therefore greatly exaggerate the disparity between the two countries.[1] While finance as well as labour was dear in America, the scarcity of finance attracted English funds more readily and earlier than the scarcity of labour drew out migrants. There were so few impediments to the import of British capital into America that the yields on long-term obligations in the two countries cannot have differed by very much more than the risk premium. An American textile manufacturer, asked by the Select Committee on Manufactures of 1833 to describe his methods of calculating capital-costs, said that he reckoned his interest upon the purchase price of the machinery and for this purpose took a rate of 6 per cent in America, and of 5 per cent in England.[2] The choice of these rates would have biased the

[1] For interest rates in the mid-West, see T. S. Berry, *Western Prices Before 1861* (Cambridge, Mass. 1943), pp. 411, 441, 513.

[2] P.P. 1833, VI, Qu. 2617. When Gallatin wished to express as an annual figure the capital expended on barns in the Middle and Northern States he used a rate of 5 per cent. 'Suggestions on the Banks and Currency of the several United States', 1841, in *The Writings of Albert Gallatin*, ed. Henry Adams (Philadelphia, 1879), III, p. 254.

American choice of technique towards the more capital-intensive methods.

In many cases, however, early manufacturers neglected to take account of interest except when they had to borrow from outsiders; each year they withdrew from the business enough to live on and ploughed back the rest irrespective of the yield on alternative methods of employing their funds. In effect, they behaved as if their capital cost them nothing. If entrepreneurs in both countries behaved in this way finance would certainly have appeared to be cheaper, in relation to labour, in America, than in England.[1]

The nature of the spectrum of techniques

The practical importance of the inducements in the U.S.A. to adopt capital-intensive methods depended on the nature of the techniques available, and the possibilities they afforded for substitution between labour and capital. There clearly were several occasions on which one technique was manifestly superior for any likely range of factor-prices, and would therefore have been the most appropriate choice in England as well as America. The new techniques for spinning which were invented in the later eighteenth century were so much more productive for all factors than the old spindle that they were the best choice at any conceivable level of wages. But there were other situations in which the possible methods of production were sufficiently competitive, one with the other, for the manufacturers' choice to have been influenced by relative factor-prices. It is difficult to say how far the various new methods of spinning were substitutes for each other—Hargreave's jenny, Arkwright's water-frame and Crompton's mule—and how far the suitability of each type of machine for the production of particular types of yarn specialised their uses; but, at some stages of development, some manufacturers may possibly have been influenced, in choosing between them, less by the market for particular grades of yarn than by the relative costs of the methods in the production of similar grades. In the years immediately after its invention the power-loom was not so decisively superior to the hand-loom that its adoption was uninfluenced by relative factor-prices; and even as late as 1819 it was not clear that in England the saving of labour

[1] For a further discussion see below, pp. 69–73, 167–8.

was sufficient to outweight the increase in capital-costs of the power-loom: '... one person cannot attend upon more than two power-looms, and it is still problematical whether this saving of labour counter-balances the expense of power and machinery, and the disadvantage of being obliged to keep an establishment of power-looms constantly at work'.[1] On balance it seems reasonable to suppose that in the textile industry in the first half of the nineteenth century, the range of possible methods of production was sufficiently wide and continuous in respect of the proportions in which they used capital and labour for the choice of techniques to be responsive to relative factor-prices. And though the point could be settled only by detailed investigation, there is a general reason for expecting that similar conditions prevailed in other industries often enough to make labour-scarcity worth considering.

In the first place technical progress was still more empirical than scientific, that is it depended more on the response to particular and immediate problems of industrial practice than on the autonomous development of scientific knowledge. Technical development was therefore likely to take the form of slow modifications of detail, as opposed to spectacular leaps to a new technique decisively superior from the start to its predecessors; most even of the 'great inventions' of the period resolve themselves on close inspection into 'a perpetual accretion of little details probably having neither beginning, completion nor definable limits'.[2] (For the same reason the process

[1] Quoted by S. J. Chapman, *The Lancashire Cotton Industry* (Manchester, 1904), p. 31. Strassmann suggests that the improvements in the first half of the nineteenth century were predominantly labour-saving and capital-using. (W. P. Strassmann, *Risks and Technological Innovation* (Ithaca, 1959), pp. 118–19.) The improvements in flax-spinning machinery in the 1830's saved labour but required additional capital per unit of output (P.P. 1841, VII, Qu. 3086–3092). A witness before the Select Committee on Manufactures, Commerce and Shipping of 1833 estimated that the efficiency of spindles had increased by 20 per cent since 1815, but that 15 per cent more machinery was required (presumably at the preparatory stage) 'to effect that improvement in quality which enables us to do that extra quantity'. This suggests that there had been no significant decrease in machine costs per unit of output of yarn over these years. (P.P. 1833, VI, Qu. 5263.) The self-actor mule, in a case quoted in 1842, had higher capital-costs per unit of output than the hand-mule. (P.P. 1842, XXII, Factory Inspectors' Reports, p. 364.) Whitney's method of small-arms production required heavy and expensive machinery which was said to be worth installing only if it could operate for at least twenty years. (J. Mirsky and A. Nevins, *The World of Eli Whitney* [New York, 1952], p. 245.) This suggests that the method was capital-intensive in our sense.

[2] S. C. Gilfillan, *The Sociology of Invention* (Chicago, 1935), p. 5.

of improvement was more likely to be sensitive to the factor-needs of the economy in which they were made.) In the second place a large sector of industry was organised on the domestic system. Under this system circulating capital was more important than fixed, and the commercial capitalist was always facing the question in what proportions to distribute his investment between fixed and circulating capital, that is, how much of his funds to lay out in raw materials to be worked up by domestic workers and how much on machines of his own. This choice was very sensitive to the cost of labour; and so long at least as the costs rose and indicated a shift *into* fixed capital, the commercial capitalist was in a better position to respond to the stimuli than his successors with a high proportion of their funds locked up in fixed capital. It was after all for an industry still organised principally on the domestic system that the Ricardo effect was postulated.

Moreover, even when the range of basic techniques likely to interest a manufacturer was very narrow or when one process was distinctly superior over a very wide range of relative factor-prices, it was possible to use them in a more capital-saving or a more labour-saving way, for example by varying the number of machines per worker, by running the machines for shorter or longer hours (by arranging workers in shifts) or at more or less rapid rates,[1] and by variations in the amount of space per worker or per machine.

The existence of methods of varying the factor-intensity of the basic techniques meant that there was usually a fairly continuous range of methods and that the method which used a little more capital saved a little more labour. But relative factor-prices would still be influential even if there were dis-

[1] A witness before the Committee on the Export of Machinery (1841), who had had experience of working in American cotton mills in the 1830's, said that cotton machinery was worked slower in the U.S.A. than in Britain, because the English employed more spindles per frame. (Qu. 1830–1835.) This was also the opinion of George Wallis in 1854: 'The general speed of power-looms, and indeed of machinery generally, is lower than in England. By this means labour is economised, and one labourer can attend to more machines'. (P.P. 1854, XXXVI, p. 21.) Since running machinery more slowly is a method of saving labour at the expense of an increase in capital per unit of output, it is what we should expect in a country of dear labour. On the other hand Montgomery, who also had a detailed knowledge of the cotton-textile industry in both countries stated categorically that the Americans ran their spinning machinery faster than the English. For a possible reconciliation of this apparent conflict of evidence see below, p. 54, footnote 2.

continuities such that, at some point, the alternative technique saved a great deal of labour but required a great deal more capital. Indeed the most striking disparities between English and American technology were probably established in just such cases. The gap between the hand-loom and the power-loom, and again, towards the end of the nineteenth century between the ordinary power-loom and the automatic loom, was wide. The capital-intensity of the 'superior' machine could be modified by running it longer, which some Americans did, but the need to make this modification is probably itself evidence that the new technique was much more capital-intensive than the old. The conditions of their labour-supply gave the Americans a much stronger incentive than the English had, to leap such a gap in the spectrum of techniques, with effects on subsequent technical progress which will be discussed later.

We have so far considered only the consequence for the choice of technique of the fact that labour was scarcer in America than in England, and have referred to natural resources only in so far as the abundance of agricultural land was a condition of the scarcity of labour. We must now take natural resources more explicity into account.

The price of natural resources had an effect on the choice of techniques ranked by reference to the proportions in which they employed capital and labour. If natural resources were employed in the same proportions in the capital- as in the labour-intensive techniques, the price of natural resources would not affect the tendency of a rise in wages to shift the manufacturers' choice towards greater capital-intensity. If the supply of natural resources were inelastic, so that attempts to widen capital met rising costs for natural resources as well as for labour, the bias towards capital-intensity would be *strengthened* if the capital-intensive technique saved natural resources as well as labour, and *weakened* if it was more expensive in natural resources.

But this does not exhaust the possible effects of natural resources. For there may have been some alternative techniques, the principal difference between which was in their possibilities of substituting between natural resources and either capital or labour or both. Some techniques were important principally because of the proportions in which they used capital and natural resources: large blast-furnaces may have allowed a substitution of capital for raw materials, and the application of steam to water-transport may have involved the reverse sort of substitution—of power for capital; it is also possible that methods of building factories differed in respect of the proportions in which they used land and capital.[1] There were also techniques in which there was substitution between natural resources and labour; in particular there were possibilities of using power from water and steam instead of man-power.

Because of the unhomogeneity of natural resources and variations in their price within regions it is impossible to make any general statement about their cost, in relation to labour and capital, in the two countries. Land was certainly more abundant in the U.S.A. in relation to both other factors, and this fact dictated the choice of American agricultural techniques, which substituted land for labour. The abundance of land, and the nature of the American climate also enabled some substitution of natural resources for capital. There was less need than in England for investment in farm-buildings—the maize-stalks were left standing in the fields and they provided winter shelter for the cattle who were sometimes not brought in at all; and because of natural pasturage there was less need for winter feed. In some regions the type of agriculture was influenced by the ability to substitute natural resources for capital and/or labour: maize growing, for example, was a labour- and capital-saving, land-intensive form of agriculture. American agricultural methods which 'mined' the soil in effect substituted natural resources for labour and capital and so did the use of wooden frame houses. In industry too the lower rents for sites enabled New Englanders to economise in labour and capital in the construction of cotton-textile mills and also to build mills which enabled more effective use to be made of the textile workers and textile machines by allowing them more space. Similarly the American railways were built in ways which, in

[1] The effect of lower rents for American industrial sites was to allow manufacturers to give greater weight to considerations which were favourable to ample factory space in all countries. 'One distinguishing feature of manufacturing establishments in the United States, both public and private, is the ample provision of workshop room, in proportion to the work therein carried on, arising in some measure from the foresight and speculative character of the proprietors, who are anxious thus to secure the capabilities for future extension, and in a greater measure with a view to securing order and systematic arrangement in the manufacture.' (P.P. 1854–5, L, p. 630).

effect, substituted land for capital, as contrasted with the English railways which were built with a disregard for natural obstacles, a disregard which increased their engineering cost.

Almost certainly also, power was cheaper, in relation to capital and labour, along the Fall Line and its supply more elastic than that available to some areas in England, and it may be that the mechanisation of the Massachusetts cotton-textile industry was a substitution not so much of capital for labour as of cheap water-power for labour. If, in order to use cheap power, it was necessary to use more capital per unit of output, the high cost of labour gave an additional inducement to the substitution, but if the power was very cheap the substitution might have been profitable even had American wages been at the English level; and in support of the argument it might be pointed out that mechanisation was much slower in sectors of the American cotton-textile industry, for example in Rhode Island, where labour was no less dear but where power was expensive. Moreover, water-power was, in effect, substituted for capital as well as labour.[1] The Americans ran certain types of textile machinery faster than the English, and this practice represented, to some extent, a substitution of natural resources for capital. 'Driving machinery at high speed', wrote Montgomery, 'does not always meet with the most favourable regard of practical men in Great Britain; because in that country where power costs so much, whatever tends to exhaust that power is a matter of some consideration; but in this country (that is, the U.S.A.), where water-power is so extensively employed, it is of much less consequence.'[2] The American cotton-textile manufacturers also obtained their raw cotton on somewhat better terms than did the English, and this enabled them to economise in labour by using a better grade of cotton; in Lancashire manufacturers economised cotton at the expense of wages, using a great deal of short-stapled cotton.[3] In the construction of ships, cheap timber enabled American ship-builders to economise in labour and capital.

[1] As was wood where it could be used to provide power. The Americans developed high-pressure locomotives partly to overcome steep ascents and partly 'on account of fuel not being so much an object'. (P.P. 1841, VII, Qu. 3000.)

[2] Montgomery, *op. cit.* p. 71.

[3] The greater need to make economical use of the raw material was one reason for the English preference for the mule-spindle, since the English used shorter-stapled cotton than the American and reworked more of the waste (Copeland, *op. cit.* p. 72).

This general line of argument is tantamount to the familiar view that the high productivity of American industrial labour was due principally to the fact that it was combined with richer natural resources, rather than with more capital, though sometimes more capital per head may have been technically necessary to combine the labour with the resources.

But whatever force this argument may have for the later nineteenth century, in the period we are discussing it is not evident that, with the possible exception of cotton and wood, the natural resources relevant to industrial manufacturing were cheaper in relation to capital and labour in America than in Britain. Outside the Fall Line the supplies of power in the U.S.A. may well have been dearer and less elastic, in relation to labour and capital, than in England, since the supplies of coal were small; and until the discovery of new sources in the 1860's and '70's the same may have been true of iron-ore. In the manufacturing districts of New England, wrote Cobden 'the factory system has been planted under great disadvantages from the dearness of coal and iron'.[1] Moreover the technical possibilities within industry of substitution between natural resources on the one hand and capital and labour on the other were less than the possibilities of substituting between capital and labour. We feel justified therefore in proceeding on the assumption that the dearness of American labour is the most fruitful point on which to concentrate in an examination of the economic influences on American technology.

Some implications of the argument

It will be seen that scarcity of labour is necessary to the argument in a number of ways. It favoured a pattern of ownership which provided the farmer with an incentive to increase productivity per head. Since the American farmer owned his own farm and his family provided the bulk of his labour he had a direct interest in increasing the value of his land with a view

[1] Cobden, *op. cit.* p. 207. 'In the sea-port towns, the European coal comes into competition very successfully with the native coal, especially the Virginia coal' (P.P. 1841, VII, Qu. 2943). Newcastle coal at New York was about £1 10*s.* 0*d.* per caldron; coal at Pittsburgh was 6 or 7 shillings a caldron, but 'if the coal has any distances to go, over 100 miles, you increase the price 100 per cent above the Pittsburg prices'. The manufacturers of machinery paid more for the raw material in America than in England, on an average 100 per cent more. (P.P. 1841, VII, Evidence of Alexander Jones.)

to its resale, and he stood to gain the entire reward for easier work and higher productivity. Quite apart from the pattern of ownership, scarcity of labour ensured that, within the limits set by geology and climate, American agriculture developed along land-intensive, labour-saving lines, that is, assumed high labour-productivity forms, and that the benefits of this high productivity accrued to the cultivator.

It was neglect of these considerations which led Ricardo to argue that the high productivity of agriculture in new countries would militate *against* mechanisation. 'Old countries', he wrote, 'are constantly compelled to employ machinery and new countries to employ labour'.[1] This conclusion was based on the assumption that the wages of labour in new countries were determined, like those of English labour, by the cost of a subsistence diet which would be lower, and would rise less for a given increase in population, where fertile land was abundant. This assumption, however, was not correct. Where labour is scarce, wages are not determined by the cost of subsistence along Malthusian lines, and the effect of high agricultural activity in the new countries was to *raise* the wage necessary to attract away from agriculture into industry.[2]

But in addition to scarcity of labour, the specific qualities of American land are also essential to the argument, as well as the political influences on American land policy. The point may be made by examining the contrast with the Argentine, another country of recent settlement with a sparse population. The Argentine in the mid-nineteenth century devoted its land to pastoral farming, the most land-intensive, labour-saving form of agriculture. This was partly because of the natural advantages of the land for this form of agriculture and partly because the Argentine was dominated, from early times, by owners of great estates. These two circumstances, of course, reinforced each other: if the land had been better suited for intensive agriculture, it would have been more difficult to maintain a system of great estates; if the country had not been dominated by its great landowners, it would have been possible

[1] Ricardo, *op. cit.* p. 41. 'In America and many other countries, where the food of man is easily provided, there is not nearly such great temptation to employ machinery as in England, where food is high, and costs much labour for its production.' (p. 395.)

[2] See W. A. Lewis, 'Economic Development with Unlimited Supplies of Labour', *The Manchester School*, XXII (May 1954).

to investigate more seriously other possibilities than pastoral farming.[1] But the net result was that the floor set to the industrial wage by openings in agriculture was a low one. In the U.S.A. on the other hand, given the qualities of the land and market conditions, the most profitable form of agriculture was cereal production, and although it was conducted by the most labour-saving methods available, it was, from its very nature, much more labour-intensive than pastoral agriculture. In the New England colonies the character of the early settlement was, of course, favourable to the development of a community of small farmers, as opposed to a society like that of the Argentine dominated by great landowners. But the more strictly economic circumstances also ensured that agriculture was conducted by small farmers and that the benefits of high agricultural productivity accrued to them.

But it was not merely high agricultural productivity and the fact that this was enjoyed by owner farmers that maintained returns in agriculture. An additional influence was exerted by the existence of a frontier, that is an area of undetermined size not yet brought into cultivation or subject to unambiguous private title. Since so much of the development in regions of recent settlement took the form of an extension of cultivation into new areas, the chance of gaining from a rise in land values was open to a relatively large number of people. In Europe the growth in population and wealth did indeed bring windfall gains to existing owners of land, but the market in land was inactive and the chance of making such gain correspondingly limited; thus though gains from rising land values in Europe undoubtedly exerted some influence, for example in maintaining the position of a section of the landed nobility in England, their effect on economic development was altogether smaller than in the regions of recent settlement.

The nature of the gains which arose on the frontiers of the nineteenth-century world varied from time to time and area to area.[2] At some times and in some regions the expansion of

[1] For an interpretation of Latin-American experience which attaches greater importance to the independent effects of social structure, see 'Economic Growth in Latin America' by B. F. Hoselitz in *Contributions to the First International Conference of Economic History* (Stockholm, 1960), pp. 87–101.

[2] On the economic effects of the movement of the frontier I have profited greatly from discussion with Dr Guido Di Tella and from an unpublished paper on 'The Economic Growth of Empty Spaces' by him and M. Zymelman. See also J. Robinson, *op. cit.* p. 294.

the frontier lagged well behind the increase in the profitability of bringing new land into cultivation. Throughout the nineteenth century improvements in transport, the evolution of new strains of crops and breeds of animals, and new methods of draining and financing reduced the costs of agriculture on the frontier. And the rise in consumers' incomes, the fall in transport-costs and the rise in the costs of alternative sources of supply raised prices to the farmer on the frontier. The expansion of the frontier did not respond smoothly to these changes in profitability. In addition to all the well-known reasons why it is particularly difficult for farmers to respond to market forces in an economically rational way, there were special circumstances which at times held back the expansion of the frontier. Sometimes the facts of geography were responsible, for example the difficulties of penetrating the Alleghenies; sometimes government policy, for example the attempt in Australia, under the influence of Wakefield's theories of colonisation, to maintain dear land within 'an iron boundary of settlement'. Sometimes the lag was due to lumpiness in the investment involved in the expansion of the frontier, for example when the cost of the minimum feasible transport-system was large in relation to existing resources, so that entrepreneurs needed to be assured of a large fund of unexploited potentialities in a region before they embarked on bringing it into use. For all such reasons the expansion of the frontier might at times be slower than the returns on the frontier warranted.

It is not argued that this was true of the frontier in all parts of the world. There may have been cases where labour and capital moved to the frontier more or less at the rate warranted by their yields on the frontier, in such a way that the extensive margin of cultivation was established, in fact as well as in the text-books, where the marginal product of land was zero. There were certainly some frontiers where expansion was too rapid from an economic point of view, for example in South Africa as a result of the Great Trek: cases where the movement from land already settled not only added nothing to the total product but reduced it. And, in many frontiers, periods when expansion lagged behind what was economically warranted were matched by other periods when expansion overshot the mark, and this was particularly likely to occur when the expansion periodically involved the competitive overbuilding of railways. All that it is intended to argue here is that at times in the first half of the nineteenth-century economic conditions, narrowly defined, warranted a faster expansion of the American frontier than took place. There is no doubt that changes in costs and prices were long-term conditions of this expansion; with the techniques, interest rates and product-prices prevailing before 1800 it would not have been profitable to open up the middle west, and the invention of the railways and/or a rise in consumers' income were necessary before it could become profitable. But it took time before entrepreneurs responded to the changes in costs and prices, before they assessed the possibilities correctly and projected and built the railways, and before farmers took up holdings.

Where such a lag existed, the margin of settlement stopped well short of the point of zero rent, and investment in extending the frontier in these conditions yielded a return in excess of the 'normal' rate of profit. Even when the extension of the frontier did proceed *pari passu* with changes in profitability, anyone who acquired property on the frontier stood to enjoy windfall gains if future changes in cost/price ratios carried the frontier beyond him; and the gains from correctly guessing the precise speed and path of development might be substantial.[1]

Not only were the unexploited potentialities greater on some frontiers than on others, but the gains from exploiting them were also distributed differently according to differences of land tenure, government policy etc. In the U.S.A. a large part of the gain accrued to the railways because they were able to resell the large tracts of land they had been granted by the state and, by charging high freight rates, were able to tap the gains

[1] Professor Curti in his study of Trempealeau County, Wisconsin, found evidence of 'phenomenal gains' among settlers in the region, particularly between 1860 and 1870. M. Curti, *The Making of an American Community* (Stanford, 1959), ch. VIII. For a recent estimate of gains on speculation in frontier lands see A. G. and M. B. Bogue, 'Profits and the Frontier Land Speculator', *Journal of Economic History*, vol. XVII, No. 1, 1957, pp. 1–24. The argument in the text is not incompatible with the existence of periodic complaints that profits were lower in agriculture than in industry; indeed we should expect such complaints in cyclical depressions which in agriculture took the form of low prices, and in industry of unemployment. But it is not compatible with the view sometimes implied that the returns to farming were persistently lower than those in industry. (See T. Saloutos, 'The Agricultural Problem and Nineteenth Century Industrialism', in *Economic Change in America*, ed. J. T. Lambie and R. V. Clemence [Pennsylvania, 1952], p. 318.) This view would make it difficult to understand why the West was ever settled.

accruing to farmers. But in some places the existence of water-transport limited the power of the railways to charge high freight rates and in other places the overbuilding of railways strengthened the bargaining position of the farmers. In these ways part of the unearned increment which arose in the course of frontier expansion accrued to the farmers who moved to the frontier. These gains played an essential part in the process by which new land was brought into cultivation. The more enterprising farmers took up holdings on the frontier and cleared them, often raising a mortgage for the purpose. As the country became more thickly populated, the value of land in the region rose, and the pioneer farmer was able to sell out to a newcomer and thus realise not only his investment in clearing the property but the increase of value due to the increase of population. For the farmer on the moving frontier the annual income from the farm was probably less important than this deferred return on his capital and effort. With it he could push on and clear another farm.

The fact that returns to agriculture in the frontier regions were sustained by these possibilities of gain had two conflicting effects. It can be argued in the first place that investment on the frontier was boosted, and that those frontiers where expansion at times lagged well behind what was warranted by cost/price ratios may in the long run have developed more rapidly than those where expansion was more continuously consistent with increases in profitability. Expansion by periodical bursts and surges which more than made up the periodic lags may have been more effective and have led to a greater increase of incomes than expansion by small increments of continuous change. If this was so the possibilities of gain were favourable not only to the expansion of agriculture, but to the growth of those types of industry developed to process the raw materials produced by the expanding agricultural sector, industries producing heavy agricultural equipment, and industries producing the simpler manufactures for the local market where transport-costs kept out imports. On the other hand the possibilities of gain had an influence on the terms on which resources were available to industry over and above the effect of the high productivity of agricultural labour. At times the comparative advantages of agriculture were exaggerated, and the pull on potential industrial labour was accentuated. The collapse of the land boom might set up a reverse movement but this was not full compensation.

Thus frontier expansion had conflicting influences on industrial development, and the balance was struck in different ways in different places. In some of the regions of recent settlement it can plausibly be argued that the net effect was to brake the development of industry. In addition to the normal troubles of infancy and external diseconomies, their industries had to face the competition of high returns in agriculture and frontier settlement. It was only when the frontier was 'closed' and the profits of agriculture ceased to be boosted in the ways we have described that industrial growth became rapid, that is, that the change in the balance of comparative advantage between land-using activities and industry came principally from a fall in the profitability of the former. What requires explanation in the case of the U.S.A. is the development of a substantial industrial sector characterised by considerable technical progress and quite a rapid rate of growth, that is, a shift in the balance of comparative advantage arising from the increasing efficiency of American industry, long before the closing of the frontier.

There is no difficulty in explaining why the methods employed in American agriculture should have become more mechanised than those in Britain. With abundant land and scarce labour the marginal product of an agricultural machine was obviously greater on the plains of Ohio than on the fields of the English Midlands; in the U.S.A. the machine enabled land to be cultivated which would otherwise not have been cultivated at all, while in England it merely replaced labour. And in England, even where the machine was so productive that it could profitably have replaced even the low-paid English labour, the abundance of labour presented social obstacles to its adoption.

But the comparative advantage of the United States at the opening of the century was so predominantly in agriculture that it is less obvious how the argument we have pursued could explain mechanisation and technical progress in *industry*. On the face of it one might expect high productivity in agriculture and abundant land to hinder the rise of industry as much as to ensure that industry, when it arose, was highly mechanised.

This question will be discussed in more detail later, but there are some points which ought to be made now. One possibility is that agricultural incomes in the U.S.A. increased more rapidly than in other primary-producing regions of recent settlement, and that the protection of distance ensured that a large part of the consequent increase in demand for manufactured goods was concentrated on local industries. For obvious geographical reasons the U.S.A. was better suited for settlement than the other regions of recent settlement, and certain sections produced crops (in particular cotton) which were in demand in the older industrial countries. The protection which distance provided to her own industries was substantial. In 1775, average charges covering land carriage, insurance and commissions, for goods shipped from North England to America via London were 32 per cent of their value, exclusive of freight. The cost of marketing within America was even higher: in the early nineteenth century wholesalers generally charged a mark-up of 100–150 per cent of the cost of imported merchandise.[1] This protection of distance was clearly of great importance; had American development been delayed—as the Australian was—until the revolutionary reductions had been made in sea transport-costs, one might conjecture that America would have concentrated much more heavily than she did on land-intensive activities.

Moreover the development of agriculture in new and more fertile parts of the U.S.A. turned the balance of advantage in the New England states against agriculture. We have so far considered the U.S.A. as a single region of recent settlement in which the returns to capital and labour were determined by their yield on the moving frontier. But for some purposes we ought to consider the New England states as a separate economic region in which the frontier was reached and the possibilities of capital gain eliminated from the returns to agriculture at an early date. The balance of advantage in favour of agriculture in the New England states had always been narrow, and the development of new lands and the opening of the Erie Canal reduced the profits of agriculture in this region. Since the internal market in agricultural products was, at least from the 1820's on, more perfect than the labour- and capital-markets, agricultural profits were reduced by the inflow of

[1] L. E. Atherton, *The Southern Country Store* (Baton Rouge, 1949), p. 170. A. H. Cole, *The American Wool Manufacture* (Cambridge, Mass. 1926), I, p. 50.

products from new areas of greater fertility, but capital and labour did not flow out at a rate sufficient to maintain wages and profits in New England. The balance of advantage of this area *vis-à-vis* the rest of the U.S.A. was shifted in favour of industry, and as industry developed it increasingly enjoyed the benefits of division of labour and economies of scale, which shifted the balance further in favour of industry. As early as 1832 New Hampshire farmers complained that factories attracted labour and raised wages so that agriculture became unprofitable.[1]

How large an industrial sector would have emerged from these influences alone—growing agricultural incomes, a shift in the balance of advantage in the New England states away from agriculture, plus the protection of distance—is anyone's guess. In industries producing goods so bulky that they did not enter into international trade, and in the processing and extractive industries, these influences may be sufficient to explain what happened. But some of the industries which developed in the U.S.A. produced goods for which distance did not provide adequate protection, and it is difficult to explain the early course of their development without attaching importance to the tariff. The tariff is necessary to the argument at two points and is a possible help on a third.

(1) Protection was necessary in the first place to ensure that, though wages in industry were measured by the returns in independent agriculture, the rate of profit in industries subject to foreign competition was broadly comparable to that in agriculture; not necessarily the same, since some people prefer industry to agriculture, but sufficiently in line to induce people of initiative to become industrial entrepreneurs.

(2) But it was also necessary that the level of protection was sufficiently stationary for long intervals in the following decades to ensure that, in the case of internationally traded products, U.S. industry could not compensate, by higher prices, for the rising cost of additions to the labour-force. It was because American industrialists were already exploiting the tariff to the full at the margin, that they were prevented by the possibility of imports from passing on in higher prices the higher cost of labour. If the rates of the American tariff had been tied to an index of labour-costs, increases in the tariff might have obviated the necessity of changes in technique.

[1] Clark, *op. cit.* I, p. 392.

(3) The tariff is also relevant in a third context: in so far as it was levied principally on imports of goods which were labour-intensive (as opposed to capital-intensive), it raised real wages at the expense of capital. A tariff on goods which were labour-intensive when produced in the U.S.A. would also have tended to shift American demand away from the goods made with labour-intensive techniques towards products with capital-intensive techniques which, after the imposition of the tariff, became relatively cheaper.[1]

LABOUR-SCARCITY AND THE RATE OF INVESTMENT

In the immediately preceding subsection we have explained how the dearness and inelasticity of supply of American industrial labour gave American manufacturers an inducement to adopt methods which were labour-productive even though they were capital-intensive. But these characteristics which explain the *composition* of investment would impose a restraint on the *rate* of investment. The shift to the more capital-intensive techniques partly offset the effect of rising labour-costs on the rate of profit, to an extent which depended on the adequacy of the existing range of techniques for the purpose. But the offset—in principle at least—could not be complete since, if the more capital-intensive method was as superior in productivity as this, it would already have been adopted in the U.S.A. before the rise in wages and in England at the lower wage-level. Once labour-costs had risen, the rate of profit would be higher with the more capital-intensive method than with the less, but it would nevertheless be lower than it was before the rise in the cost of labour, and, on reasonable assumptions, lower than in England; that is, other things being equal, the inducement to expand capacity would now be less than formerly and less than in England. Moreover, unless we assume (as was obviously not the case) that wage-earners were prepared to save the full increase in their earnings, the amount of new capacity that could be financed was reduced. So long as the volume of investment depended on the profit-rate, investment in the U.S.A. must have grown more slowly than previously for, since the machines in the more mechanised methods were more expensive, in terms of output, than the simpler machines, it must now have taken longer for the American manufacturers to accumulate the capital to produce a given output; and on these assumptions

[1] It is possible, of course, that the principal effect of the tariff was to raise the real return to both labour and capital at the expense of land.

investment would also have grown more slowly than in England. In so far as the more capital-intensive equipment employed in the U.S.A. was produced in the U.S.A. the move to the capital-intensive end of the spectrum lowered the rate at which capacity was expanded to an even greater degree, because the capital-goods industry had to devote itself to producing equipment which was expensive in terms of output. To this extent labour-scarcity was a disadvantage which might partially be offset by substituting machinery, so far as this was technically possible within the range of existing methods, but which was none the less a disadvantage.

To put the matter in slightly different terms, the desire to widen capital was more likely in the U.S.A. than in England to run up against rising labour-costs. Hence the desire to widen capital was more likely to lead to its deepening in the U.S.A. than in England, deepening being the method of partially offsetting the fall in marginal profit-rates. But since, within the range of known techniques, the fall could not be completely offset, it imposed upon the desire of American manufacturers to widen capital a restraint to which English manufacturers were not subject.

This is the sort of situation envisaged by Marx. According to him, labour-scarcity—the exhaustion of the reserve army of labour—would lead the capitalist to substitute machinery for labour, that is constant for variable capital; this would lead to a decline in the rate of profit, a fall in accumulation and in the demand for labour and a consequent replenishing of the supply of labour. The size of the reserve army was maintained by variations in the total of accumulation and in the proportion between fixed and circulating capital. With a smaller reserve army of labour, the tendency to substitute machinery for labour was stronger in America; but by the same token, the rate of investment was subject to a more severe restraint.

In some industrial activities the dearness and inelasticity of labour did in fact exercise a powerful restraint on the rate of investment. This happened even in an industry like the manufacture of small arms where techniques were available which, at first sight, might be supposed adequate to compensate for the high cost of labour. 'High wages', wrote the Chief of Ordnance in a report to the Secretary of War in 1817, 'makes the business unprofitable to the contractors, and ultimately in many cases has occasioned their ruin'.[1] The true rate of profit on the manufacture of arms, when the capital-costs were accurately accounted, was low; and many concerns remained in business only because their primitive accounting concealed the fact that they were, in effect, treating capital as income and failing to provide for depreciation. It has been said of Simeon North, one of the inventors of the method of interchangeable parts, that: 'through the withdrawal as profits of sums which should have gone to pay for renewal charges, he squeezed his factory dry of its productive capacity and was forced after some years to start over again with new investment'.[2] For these, among other reasons, a very high proportion of the early small-arms manufacturers went into liquidation.

Nevertheless in a number of industries investment was rapid and in the industrial sector as a whole it is at least not evident that investment was slower in America in the first half of the nineteenth century than in England, despite the restraints of dear labour. It may be that the assumption which creates a problem out of this—the dependence of investment on the rate of profit on capital—is not valid; but for the moment we shall retain it, and try to consider what circumstances in America might have exerted a favourable influence on the rate of profit.

One possibility is that natural resources were cheaper and their supply more elastic in the U.S.A. than in England to an extent which offset the effects of its dearer labour. In this case the restraint which labour imposed upon accumulation in America would have been matched by a natural-resource restraint in England. But, as has already been argued, it is not evident that the natural resources most relevant to manufacturing industries were in fact cheaper in the U.S.A.;[3] and they would have had to have been considerably cheaper to offset the dear labour, since in most industrial products labour was a higher proportion of total costs than natural resources. We must therefore enquire in what ways dearness and scarcity of American labour might have favourably influenced the *rate* of investment. We shall consider three main ways.

In the first place, the American manufacturers had a greater inducement to organise their labour efficiently. The dearness of American labour gave manufacturers an inducement to

[1] Deyrup, *op. cit.* p. 48. [2] Deyrup, *op. cit.* p. 54. [3] See pp. 31–34 above.

increase its marginal productivity in all possible ways, and not merely in ways which involved the adoption of more capital-intensive techniques. The shortage of labour in America from colonial times encouraged prudence and economy in its use—Washington, for example, calculated with care the proper output of various types of labour on his plantation.[1] Americans from early times were often faced with a situation where a job had to be done—a house built or a river bridged—with the labour available on the spot, because the place was isolated and it was impossible to attract more labour. This gave them an enormous incentive to use their labour to most advantage, to make use of mechanical aids where this was possible, but in any case to organise the labour most effectively. Possibly lack of domestic servants led to an early rationalisation of domestic duties and a corresponding increase in family efficiency; certainly the shortage of labour led generally to longer hours of work, to a general emphasis on the saving of time and a sense of urgency about getting the job done. In his account of his visit to America in 1818 Cobbett observed that 'the expense of labour . . . is not nearly so great as in England in proportion to the amount of the produce of a farm'.[2] The greater productivity of America, compared with British agricultural labour was partly the result of the fertility of the land; since labour was scarce, land which yielded a low return per unit of labour was just left uncultivated. It may also have been due to the avoidance of the more labour-intensive crops (for example dairy produce) as well as to the avoidance of labour-intensive soils. The superior physique and education of the Americans may partly have been responsible. Probably also, even in 1818, the American cultivator not only co-operated with superior natural resources, but had superior equipment. But Cobbett seems to suggest that the high productivity of American agricultural labour was in some measure due to the fact that its operations were more efficiently organised. The use of labour in English agriculture was much more wasteful than in English industry, partly from inertia and habit, partly because farm labour was so easy to get.

This labour- and time-saving pattern of behaviour was established on the farm from the early days of settlement—it

[1] R. B. Morris, *Government and Labour in Early America* (Columbia, 1946), pp. 38–9.
[2] W. Cobbett, *A Year's Residence in the United States of America* (1818), p. 320.

was an ingrained attitude and not simply an economic calculation—and it was carried over into other activities. 'In England' observed an English visitor to America in 1851 'we cover our (railway) lines over with superintendents, police, guards, porters and a host of other officials; and relieve the passenger of many of those troubles which, in America, he contends with himself.' 'The American omnibus' wrote the same author 'cannot afford the surplus labour of a conductor. The driver has entire charge of the machine; he drives; opens and shuts, or "fixes" the door; takes the money; exhorts the passengers to be "smart", all by himself—yet he never quits his box.'[1] This attitude to labour was also carried over into industry and led to the more efficient organisation of operations. H. C. Carey argued that the use of female labour in the American cotton-textile industry represented a more efficient use of the labour-force than was to be found in England, where men were used for jobs that physically could be done by women if those women were given the right sort of equipment. In the U.S.A. the proportion of employed females to males was higher than in England—'women being employed *here* (that is in the U.S.A.) because everything is done to render labour productive, while *there* (in England) a large portion of the power of the male operatives is wasted'.[2] The most conspicuous example of efficient use of labour is the training that the American manufacturers gave their workers so that each was able to handle more looms.[3] Whereas, in England, the weaver spent some of his time doing unskilled ancillary jobs, the American weaver did nothing but weave. The American arrangement probably involved a somewhat lower output per loom, that is, an increase

[1] E. W. Wakin, *A Trip to the United States* (1852), pp. 130, 139.

[2] H. C. Carey, *Essay on the Rate of Wages* (1835), p. 72.

[3] P.P. 1833, VI, Qu. 5000. See T. M. Young, *The American Cotton Industry* (1901), p. 130. 'It is often found that a weaver will attend to four looms in the United States, who, in the same quality of work, would attend to only two in England.' [P.P. 1854, XXXVI, p. 21.] In 1860 the average was 4 per weaver in the U.S.A. as against 2 in Britain. In the 1880's the number of looms per weaver was 2 or, rarely, 3 in Germany, 4 in Lancashire, 6 (and sometimes 8 though at lower speeds) in Massachusetts. (Schulze-Gaevernitz, *op. cit.* p. 66 note. Copeland, *op. cit.* p. 10.) These differences partly reflect the nature of the product: more looms could be watched in the U.S.A. because there was a large demand for simple cloth, and the demand for the more elaborate, labour-intensive textiles was met by imports. Thus the differences in looms per worker was in part simply an international division of labour; but it also represents, and probably to a greater degree, the adoption in America of more capital-intensive methods of making similar products.

in capital per unit of output, but there is little doubt that the English manufacturer would have found it profitable to adopt the same method of economising labour. The point was that his need to do so was less; abundant labour, like the salt on the edge of the plate, tends to be wasted. 'Such a state of society where, as with us', wrote the author of an English text-book on weaving in 1846, 'labour generally exceeds the demand for it, has a tendency to beget indifference to its improvement'.[1] In the manufacture of small arms, also, specialisation of labour was carried much further in England and America, even before there were significant differences in technical processes; in England a workman specialised on one part of the weapons, but carried out all the operations on that part—in the U.S.A. several workmen each performed only one or two operations on the part.[2]

Even, therefore, where there were no differences in technology or at least only such as involved the different disposition of identical machines, differences in the organisation of operations may have ensured that the intensity or effectiveness of an hours' labour was greater in the U.S.A. than in England, and this tended to make up for the fact that the price of an hour's labour was higher in the U.S.A.[3] It is not an accident that scientific systems of labour management originated in America. Not only did mechanisation and standardisation make it easier to assess the effort needed for a particular operation, but the scarcity of labour made such assessment more necessary. In its turn careful management of labour bred careful habits in the worker; where labour was abundant it was wastefully used, and where it was wastefully used it was difficult for the worker to acquire the industrial virtues.

Secondly, dear labour not only provided an incentive to organise it more efficiently. It compelled American manufacturers to make a more careful and systematic investigation of the possibilities of the more capital-intensive of existing techniques.[4] Thus labour-scarcity could have had a favourable

[1] George White, *A Practical Treatise on Weaving by Hand and Power Looms* (Glasgow, 1846), p. 331.

[2] Deyrup, *op. cit.* p. 91.

[3] In civil employment 'we perform the same labour with a much less number of persons, whether officers or clerks, than in France'. (Gallatin, *op. cit.* p. 472.)

[4] The Committee on Machinery commented on 'the dissatisfaction frequently expressed in America with regard to present attainment in the manufacture and application of labour-saving machinery, and the avidity with which any new idea is laid hold of, and improved upon'. (P.P. 1854–5, L, pp. 630–1.)

effect on the rate of investment by inducing the Americans to adopt, earlier and more extensively than the British, mechanical methods which would have been the most profitable choice even at the lower wages prevailing in England.

Labour-scarcity might, in the third place, have stimulated technical progress. Technical progress, that is movements of the technical spectrum as opposed to movements along it, would, by increasing manufacturing productivity, raise or at least keep up profit-rates, whether the progress was manna from heaven or induced by rising labour-costs. But manna from heaven one would expect to drop more readily in England, since England initially had much larger supplies of technical knowledge. The point about labour-scarcity is that it constituted a favourable influence on technical progress which was exerted more strongly in the U.S.A. than in England. Any manufacturer had an inducement to adopt new methods which made a substantial reduction of cost for all factors. But in their early stages, many of the methods devised in the nineteenth century could not be confidently assumed to effect such a reduction: before they had been tried out in practice for some time, estimates of their costs were highly conjectural. Where the best guess that could be made of a new method was that it promised a reduction of labour but some increase of capital, the Americans had a sharper incentive than the English to explore its possibilities. This is to say that labour-scarcity encouraged not only a careful and systematic investigation of the costs of the more capital-intensive of existing techniques, but the early adoption of any additions at the capital-intensive end which resulted from inventions of purely autonomous origin, even when they were made outside the U.S.A. Friedrich List wrote in the 1820's after a stay in America: 'Everything new is quickly introduced here, and all the latest inventions. There is no clinging to old ways, the moment an American hears the word "invention" he pricks up his ears.'[1] Montgomery, writing in the next decade about the cotton-textile industry, considered that though the number of specific inventions originating in the U.S.A. was not high compared with those that came from Britain, the common stock of inventions was very rapidly integrated into the American economy.

Labour-scarcity also gave Americans an incentive, not only

[1] M. E. Hirst, *Life of Friedrich List* (1909), p. 35.

to explore the labour-saving possibilities of autonomous inventions, but to attempt to invent new methods specifically to save labour. And if, as we shall argue later, the technical possibilities were richest at the capital-intensive end of the spectrum, the American was likely also to be better placed to make advances wherever, for any operation, he employed more capital-intensive methods than the English; that is the composition of American investment might have had a favourable effect upon its rate.

In the early decades of the century the principal effect of labour-scarcity in America was probably to induce American manufacturers to adopt labour-saving methods invented in other countries earlier and more extensively than they were adopted in their country of origin. The number of autonomous inventions was greater in the older industrial countries. But where their principal advantage was that they were labour-saving, they were more quickly adopted in the U.S.A. and labour-scarcity then induced further improvements, each additional improvement being perhaps small in relation to the original invention. And already in the early nineteenth century there were a number of important American inventions induced directly by the search for labour-saving methods and these became increasingly common as time went on.

Moreover, it was probably also easier for the Americans to adopt such methods. In England, where labour was abundant, labour-saving was likely to involve replacing, by a machine, labour that was already employed; in the U.S.A. it involved making a physically limited labour-force more effective by giving it machinery, but without displacing anyone, and with some increase in wages. There was, therefore, less opposition in America to the introduction of labour-saving practices and machines and of administrative methods for economising labour: the fear of unemployment was less and the likelihood greater of gaining in higher wages from the increased productivity. In England, where there was a superabundant supply of hands and therefore 'a proportionate difficulty in obtaining remunerative employment, the working classes have less sympathy with the progress of invention.'[1]

[1] P.P. 1854, XXXVI, p. 146. In the United States 'the workmen hail with satisfaction all mechanical improvements, the importance and value of which, as releasing them from the drudgery of unskilled labour, they are enabled by educa-

For the same reasons, more changes in production methods came spontaneously from the workers in America than in England;[1] particularly when the worker had been self-employed earlier in life, and most of all when he had been a farmer, for he carried over into industry the inclination to seek his own methods of doing his job better. Thus in American canal-digging, the English methods were modified by the American farmers who devised a sort of primitive, horse-drawn bulldozer, similar to a device some of them had improvised on their farms. No improvement originated among the Irish navvies who dug the English canals.

If the methods adopted or developed by the Americans did no more than offset the initial disadvantage of high labour-costs, American entrepreneurs would have been on an equality with the English. In most cases, the methods must have done less than this. But in some cases they may well have done more. In exploring the borderland of blue prints, designs and embryonic ideas and hunches which lay beyond the end of the spectrum of existing techniques, it would not be surprising if the Americans hit upon some new methods which were so productive that they more than offset the high cost of labour, methods which reduced labour and capital per unit of output so greatly that they would have been the most profitable techniques even in the case of abundant labour. Very often the substantial reductions in cost came from ancillary developments and modifications made after the new technique had been operating for some time,[2] and these benefits accrued most fully to those who had adopted the method earliest; and the

cation to understand and appreciate'. For the attitude of English labour see below, pp. 142—4.

[1] 'Every workman seems to be continually devising some new thing to assist him in his work, and there being a strong desire, both with masters and workmen all through the New England States, to be "posted up", in every new improvement, they seem to be much better acquainted with each other all through the trade than is the case in England.' (P.P. 1854-5, L, p. 38.) The Americans invented an excavating machine worked by steam, which cost £500 to £1,000, ran at £4 a day and did the work of 80 men. (P.P. 1841, VII, Qu. 3024.)

[2] 'in fact it almost always happens that the inventions which ultimately come to be of great public value were scarcely worth anything in the crude state in which they emerged from secrecy; but by the subsequent application of skill, capital, and by well directed exertions of the labour of a number of inferior artisans and practicians, the crude inventions are with great time, exertion and expense, brought to bear to the benefit of the community'. John Farey, a prominent patent engineer, before the S.C. on the Laws relative to Patents for Invention. (P.P. 1829, III, p. 547.)

process tended to be cumulative, since the successful application of machinery to one field of activity stimulated its application to another, and the accumulation of knowledge and skill made it easier to solve technical problems and sense out the points where the potentialities of further technical progress were brightest.

Furthermore, quite apart from the effect of labour-scarcity on the incentive and ability to develop superior methods, the shift of American industry towards the more capital-intensive techniques provided the American machine-making industry with an active market which stimulated inventive ability among the manufacturers of machines and machine tools and perhaps also afforded it some advantages of scale. Ability to produce a labour-saving machine in one field also made it easier to develop machines in other fields. Thus the United States developed the typewriter, not simply because in America 'copying clerks could not be bought for a pittance' but also because in Remingtons, the Illinois gunmakers, there were manufacturers available who could put ideas into practical effect.[1] Standardisation could be applied not only to final products but to the machines which produced them. 'Wood machines' wrote an English expert 'are made in America at this time like boots and shoes, or shovels and hatchets. You do not, as in most other countries, prepare a specification of what you need . . . but must take what is made for the general market'.[2] For these reasons there were cost-reducing improvements in the production of machines. Certainly by the middle decades of the nineteenth century there were some fields where the cost of the superior machines, relative to that of simpler machines, was lower in the U.S.A. than in Britain, and this was an independent stimulus to the adoption of more mechanised techniques in the U.S.A. There were also fields in which a superior machine was available for some operations in the U.S.A. but not in England.

Once a number of industries had been established in the U.S.A., a rise in real wages in any one of them due to technical progress exerted a similar effect on choice of methods as the initial high earnings in American agriculture. Where labour is

[1] J. H. Clapham, *An Economic History of Modern Britain* (Cambridge, 1938), III, p. 193.

[2] J. Richards, *Wood-working Factories and Machinery* (1873), p. 171.

scarce, any increase in productivity and real wages in one sector threatens to attract labour from other industries which have either to contract their operations or install new equipment which will raise their productivity sufficiently to enable them to retain their labour-supply.

In England where labour-supplies were abundant the technical progress in a single industry was not likely to stimulate technical progress in other industries by threatening their labour-supplies. It might, of course, stimulate technical progress in other industries by threatening their markets and in some cases their supplies of raw material; but not by threatening to draw off their labour. Any tendency for wage-earners within the technically progressive industry to establish a claim upon the fruits of their increased productivity was inhibited by the existence of a reserve army of labour, and the benefits of technical progress were likely to be diffused by means of lower prices over consumers as a whole, as in the case of the English cotton-textile industry.

In these ways the scarcity of labour gave Americans a keener incentive than the English had, to make inventions which saved labour. But it also gave them some reasons for being concerned with capital-saving. Throughout the previous argument the assumption has been made that the scarcity of labour biased the American entrepreneur's search for new methods towards those which specifically saved labour; since this is what contemporaries seem to assert.[1] But scarcity of labour, by exerting pressure on profits, did also provide some incentive to search for ways of economising other factors as well. Contemporaries only rarely suggested that dear labour was a reason for saving

[1] 'A striking peculiarity in the drawing frames of this country (the U.S.A.) viz. their self-acting stop-motion, so far as I am aware, has not yet been introduced into the factories of Great Britain, nor do I believe it necessary that it should; because the helps in that country are very different from those in this. Here they are constantly changing, old hands going away, and new ones learning. . . In consequence of this continual changing, there are always great numbers of inexperienced hands in every factory; and as the drawing process requires the utmost care and attention to make correct work as well as to prevent waste, it is necessary to have the most expert and experienced hands attending the drawing frames; but this cannot always be obtained in this country as in Great Britain; hence it is more necessary to have some contrivance connected with the machinery here which will . . . prevent the work from being injured by inexperience on the part of attendants.' 'In Great Britain, where there is always a command of experienced hands, the introduction of this stop motion would be attended with no advantage.' (Montgomery, *op. cit.* pp. 57, 59.)

capital, but Montgomery seems to have been arguing in this way when he wrote: 'the expense of labour being much greater in this country (the U.S.A.) than in Great Britain, the American manufacturers can only compete successfully with the British by producing a greater quantity of goods in a given time; hence any machine that admits of being driven at a higher speed, even though it should exhaust the power, if it does not injure the work, will meet with a more favourable reception in this country than in Great Britain.'[1]

There is another link between labour-scarcity and attempts to save capital. When from a given range of techniques, the American choice was more capital-intensive than the British, this in itself provided the American entrepreneur with an incentive to reduce capital-costs, in order to modify the large amount of machinery per operative; particularly where there were indivisibilities in the equipment, he needed to get more out of his machines in a given period of time in order to bridge gaps in the spectrum of techniques. He could do this, without any significant change in the technical characteristics of the machine, by running it longer and faster. Both these were methods of paying for the machine in a shorter period of time, that is of diminishing the interest bill on the cost of the machine and increasing the speed at which the amortisation fund was built up. Because of the capital-intensity of their output, the Americans saved more on interest charges per unit of output than the English would have. (Thus, though running machines faster and longer is in effect substituting labour for machinery, it is usually a sign of a capital-intensive technique.) Montgomery observed in the 1830's that the Americans ran their cotton-textile factories longer hours than the English and drove their machinery at a higher speed 'from which they produce a much greater quantity of work'.[2] At the end of the century it was

[1] Montgomery, *op. cit.* p. 138.

[2] Montgomery, *op. cit.* p. 138. Two opinions have already been quoted to the effect that the Americans ran their machines more slowly than did the British (see p. 30 above). Montgomery himself said that no carding machines in America were driven at so high a speed as in England, and that generally they were driven at only half the speed. (*Op. cit.* pp. 32, 39.) In the passage quoted in the text above Montgomery seems to be referring to spinning machinery, though he does not explicitly limit his observation to this operation, and elsewhere (p. 162) he concluded that 'The factories of Lowell produce a greater quantity of yarn and cloth from each spindle and loom (in a given time) than is produced in any other factories without exception in the world'. It was certainly the general impression

said of the American ironmaster that he 'wears out his furnaces much faster than the English ironmaster—in America furnaces require lining about every five years—and argues that the saving of interest on his fixed capital account justifies him in so doing'.[1]

But running machines faster and longer was only one of the ways of reducing capital-costs per unit of output. When the capital-intensive labour-saving machines had been installed, there usually proved to be possibilities of technical improvement in their construction and use. For the economy as a whole, one important form of capital-saving consisted of labour-saving improvements in the manufacture of machines, and to such improvements we can apply the previous argument about labour-saving improvements in general. The cost of machines in terms of output could also be reduced by improvements which increased their performance. Many of the inventions which were capital-saving in this sense were made as a result of attempts to improve machines whose principal advantage when they were first introduced was that they were labour-saving. In the textile industries there were few specifically capital-saving inventions. The initial effect of most of the great inventions was to save labour per unit of output at the expense of some increase in capital or at least without much saving. The saving of capital came later from such improvements as the increase in the number of spindles on each mule and the increase in the speed of the spindle. It was the manufacturers who installed the more complicated capital-intensive techniques

later in the century that the Americans ran their spinning machines faster. Mule for mule, said Young, New England produced more than Lancashire. One reason for the higher American speeds in spinning was the low marginal cost of power in the New England mills. But this was not the only reason. Spinning was a branch of the industry where American equipment was more capital-intensive than the English, and the Americans had therefore an incentive to run their machines at higher speeds, so long as by so doing they did not add disproportionately to labour-costs per unit of output. The effects of higher speed on labour-costs depended on the nature of the operations. When equipment was run faster and longer, more of certain types of labour were required per unit of output, for example the amount of piecing rose with the speed of looms, and the labour-cost of repairs increased; inputs of other types of labour remained the same, and of certain types fell. The proportions of these types varied according to industry. In processes where labour-costs rose rapidly with the speed of equipment, the Americans would not have much inducement to run their equipment rapidly. But it is conceivable that in some processes the incidental, net effect of speeding-up was to reduce labour- as well as capital-costs.

[1] S. J. Chapman, *Work and Wages; Part I, Foreign Competition* (1904), p. 87. The Americans also used larger furnaces.

who were in the most favourable position to make the subsequent improvements of this type.

There were other reasons why the Americans should have been anxious to save capital and in a favourable position to do so. In textiles, for example, American machines could be run faster because the marginal costs of water-power on the Fall Line rose less sharply than those of steam in Lancashire. Possibly too Americans wanted to get their money back sooner because they were readier than the English to assume that better methods would soon be available. These reasons will be considered later.[1] At this stage the point is only that labour-scarcity could lead to capital- as well as labour-saving; Britain on balance may have had stronger reasons for wanting to saving capital, but America had some reasons which were not present to the same degree in Britain.

American manufacturers were readier than the English to scrap existing equipment and replace it by new, and they therefore had more opportunities of taking advantage of technical progress and acquiring know-how. This is a convenient place to consider the relationship of labour-scarcity to this American habit. In its extreme form the readiness to scrap is represented by Henry Ford who is reputed to have said that he scrapped existing machines whenever a new one was invented. But to judge from scattered instances and contemporary comment, this readiness was a characteristic of American industry very much earlier than Ford. The Secretary of the Treasury reported in 1832 that the garrets and outhouses of most textile mills were crowded with discarded machinery. One Rhode Island mill built in 1813 had by 1827 scrapped and replaced every original machine.[2] It would be difficult to parallel this from Lancashire. The English inclination was to repair rather than to scrap, and to introduce improvements gradually by modifications to the existing machines. John Marshall the Leeds flax manufacturer said in 1833 that his concern had 'reconstructed' its machinery twice in the forty-five years or so that he had been in business.[3] In the English textile industry

[1] This point is discussed more fully on pp. 85–90.

[2] Strassmann, *op. cit.* p. 87. For evidence of rapid replacement see Clark, *op. cit.* I, p. 370, and Caroline Ware, *The Early New England Cotton Manufacture* (New York, 1931), pp. 135–6. For scrapping in the iron and steel industry see Burn, *op. cit.* p. 187.

[3] P.P. 1833, VI, Qu. 2455.

as a whole it is doubtful whether equipment was often scrapped except when a firm went bankrupt. The American readiness to scrap was noticed in other industries. One of the first English handbooks on woodworking-machines observed that 'there are throughout American factories but few wood-machines that have been running for ten years, and if any such exist there is a good and sufficient reason for abandoning them. The life of most machines used in joinery is not on an average more than six years . . .'[1] Contemporaries seem agreed on the general pattern of American behaviour: the American got something going, obtained his profits as quickly as possible, improved upon his original plant and then scrapped it for something better. The problem is to explain his behaviour.

Scrapping is justified on strict economic grounds when the total costs of a new technique are lower than the prime costs of the old. The effect of high wages on decisions to scrap depended on the type of equipment being used and its age. The more mechanised and the newer American equipment was in relation to the English, the smaller incentive the Americans had to scrap in favour of a given new technique. This however does not exhaust the effects of dear labour on the incentive to scrap.

In the first place for a variety of reasons which we have already mentioned, the Americans tended to run machines faster and longer hours than the English; they also built them in a more makeshift fashion. In these circumstances, though capital per unit of output would still be higher in the U.S.A. than in England (otherwise it would have paid the English to adopt the same methods), the capital would be physically used up in a shorter period of time, and the American manufacturer would be in a better position to buy a new machine embodying the latest technique.

In the second place, American manufacturers seem to have expected a higher rate of technical obsolescence than the English. An American friend of de Tocqueville told him in 1832: 'there is a feeling among us about everything which prevents us aiming at permanence; there reigns in America a popular and universal belief in the progress of the human spirit. We are always expecting an improvement to be found in everything'.[2] While a rapid rate of achieved technical progress

[1] Richards, *Treatise*, *op. cit.* p. 34. [2] de Tocqueville, *op. cit.* p. 111.

is favourable to scrapping, expectations about technical progress have a more complicated effect. If the entrepreneur expects technical progress to be rapid, especially if he expects it to be more rapid than it has been in the past, if, that is, he assumes that the latest available technique will have a very short economic life-time, its high average costs may prevent its adoption. In these circumstances the entrepreneur will put off the decision to scrap until a major technical advance appears, unless he believes that, in order to acquire the experience to take advantage of such a major advance, he must keep up with all the intermediate stages.

But the expectation of rapid technical progress had other influences which were more favourable to scrapping. The entrepreneur who expected new equipment to become obsolete, not so soon as to deter him from installing it, but sufficiently soon for him to want to ensure against the possibility, would not pay for durability. This is another reason for the flimsiness of much of American equipment. The American friend of de Tocqueville who has already been quoted said 'I asked our steamboat-builders on the North Bank a few years ago, why they made their vessels so weak. They answered that perhaps they might even last too long, because the art of steam navigation was making daily progress. In fact, these boats which made 8 or 9 knots could not, a little time afterwards, compete with others whose construction allowed them to make 12 to 15 knots.' The less optimistic expectations about technical progress among the English is one reason for the durability and heaviness of English machinery; no doubt the professional pride of machine-makers was mainly responsible, but if their customers had calculated on a rapid rate of technical obsolescence they would surely have been able to modify the prejudices of the engineers. There is another point. Because the Americans made their arrangements on the assumption that better methods would soon be available, they were more concerned to get back their money on new equipment as soon as possible. At the end of the century a French delegation to the Chicago Exhibition reported that American manufacturers invariably seemed to amortise their capital with the settled intention of replacing their machines by new and improved patterns.[1] And this was probably the reason why earlier in the century Americans had

[1] E. Levasseur, *op. cit.* p. 61.

the reputation of wanting their profits quickly. Thus technical, economic and physical obsolescence were more likely to coincide in the U.S.A. than in England at least in those branches of activity in which American expectations about the rate of technical progress were most closely realised.

In the third place, in so far as technical progress took the form of inventing machines which saved labour, but with some increase (or at least no demonstrable saving) in capital-costs it might pay the Americans to scrap when it did not pay the English. The same considerations which warranted the American manufacturer shifting towards the capital-intensive end of the spectrum of existing techniques when he was adding to his equipment, might also warrant his replacing existing equipment when new methods were invented at the capital-intensive end. Where labour was abundant, and widening of capital could proceed at a constant wage, there was no inducement to replace existing equipment unless the new equipment yielded a higher rate of profit on the value of the old and new machines together. Where labour was scarce the preoccupation of the industrialist was with retaining or expanding his labour-force. His primary interest was with methods which would increase the productivity of labour and this was a more urgent concern than the return on capital, at least in the short run and so long as the return was enough to service any external finance and provide a conventional minimum return to the manufacturer. Accounting methods in the early nineteenth century were primitive—it was easier to calculate the likely labour-saving of a new process than its capital-costs. Manufacturers had to make their choice of technique on very rough-and-ready calculations on extremely inadequate data. The bias imported into the calculations by the nature of the labour-supply could therefore be the decisive factor. The American manufacturer was averse to retaining old equipment when more labour-productive equipment was available because the old equipment made poor use of his scarce labour. So long as the saving of labour was vouched for, the capital-costs were less important, at least within a fairly wide range, and in the absence of clear ideas and relevant data about the proper components of capital-costs, manufacturers were probably disposed to underestimate rather than overestimate them. But where, as in England, labour was abundant, and there was no pressing *need* to scrap, the calculations *had* to show, in order to warrant scrapping, a *higher* rate of profit on both machines than on the old equipment, and the results of calculations in these terms was almost inevitably biased against scrapping. The crucial difference is where the onus of proof rested: in America the presumption was in favour of any equipment which raised labour productivity; in England the presumption was in favour of existing equipment—the onus of proof was on the new equipment, it had to be demonstrated that it would yield a higher rate of profit.

The fact that maintenance-costs were mainly labour-costs, and that they tended to increase rapidly with the age of equipment, reinforced the American inducement to scrap; the costs of keeping a given piece of equipment intact were greater than in England.[1]

Given the high costs of labour, and the inadequacy of existing accounting concepts and data, the readiness of Americans to give new labour-saving methods the benefit of any doubt about their capital-costs was a rational one. Even so, in particular cases, it may have led to scrapping in circumstances when it was not justified; the scrapping of old machines and the installation of new ones must sometimes have involved wastes which only capitalists who enjoyed superiority in respect of some other factor could afford to bear. It has been suggested for example, with particular reference to the replacement of horse-drawn trams by electric trams, that the greater readiness in America than in Europe to discard equipment may have been due to an inadequate analysis of the costs of change.[2]

But American readiness to scrap was partly unrational from an economic point of view. The expectation of more rapid technical progress was to some extent the result of the general optimism of the American character, and was initially independent of economic facts, rigorously defined. The Americans were, as Cobden said, 'a novelty-loving people'. In so far as the decision to scrap was taken from mere love of the latest method, it was even more likely to let down those who acted on it. Even in such cases, however, the decision, though 'un-

[1] An English machine manufacturer, in evidence before the R.C. on Export of Machinery (1841) said that repair required more skill than manufacture (Qu. 3112).

[2] H. Jerome, *Mechanisation in Industry* (Nat. Bureau of Economic Research, 27; New York, 1934), p. 333.

rational', may have turned out to be warranted by the eventual course of technical progress. Ford's readiness to scrap whenever a new machine was available might be reconciled with the orthodox criteria by assuming that, though it could not be reliably predicted of the new machines *before* their introduction that they would reduce costs to the requisite to justify scrapping, in more cases than not they proved to do so in practice. De Tocqueville's informant claimed that this was true of American ship-building, for he continued, 'And in fact this (that is, the expectation of improvement in everything) is often correct'.[1] Possibly also—a variant of the same explanation—the constant pursuit of the latest invention may have led to the adoption of machines which, though not themselves economic, ultimately put those who adopted them in a position to take advantage of later inventions which made spectacular reductions in cost. If, as is argued later, the more capital-intensive of the existing range of techniques had the greatest possibilities of technical progress, a persistently optimistic view of the costs of new labour-saving methods led in the long run to the accumulation of experience and to further technical progress which outweighed the waste of capital into which it sometimes led American entrepreneurs.

This technical progress was not without its costs. The pursuit of the latest method must sometimes have dissipated a firm's resources without yielding a commensurate increase in experience. Possibly there was too much imitation of the successful leaders, and capital-intensive methods were adopted by some concerns which would have been better advised to close down. Expectations about technical progress were not always realised, and equipment which could have been made more durable at modest expense when first installed had to be renewed at greater expense because it had fallen to bits well before new methods became available. Sometimes equipment was so makeshift that it quickly deteriorated, worked very inefficiently and was costly in repairs. But it seems reasonably clear that on balance and over the economy as a whole, the American habit of not building the equipment to last, and the closely-associated readiness to scrap were favourable to growth. For they meant that the American capital stock tended to be younger than the English and to embody more technical

[1] de Tocqueville, *op. cit.* p. 111.

knowledge. Once achieved technical progress in any line became more rapid in America than in England, this in itself weighed strongly in favour of earlier scrapping; the point we have now been making is that the American propensity to scrap developed *before* technical progress was more rapid in America, and is therefore to be included among the independent sources of such progress. In this field, if not in others, what entrepreneurs expected came to pass because they expected it.

The scarcity of labour may therefore have exerted favourable influences on the rate of investment by inducing (i) a more efficient organisation of labour, (ii) a more rapid adoption of autonomous inventions, (iii) a higher rate of technical progress and (iv) greater readiness to scrap existing equipment in order to take advantage of technical progress. There is also another way in which the composition of American investment might have exerted a favourable effect on its rate. The propensity to save out of profits may have been a function of the degree of capital-intensity. A plausible case can be made out for this view. A man who runs a machine is more likely to be interested in the possibilities of mechanisation than the man who runs a sweat shop. The industrial capitalist whose capital was tied up in his factory and its equipment was more concerned with technical development than the commercial capitalist under the domestic system—and indeed the gain, in the early days, from any shift from the domestic system to the factory may have arisen principally from the increasing role it assigned to the type of capitalist who was interested in technical possibilities. 'The threat of obsolescence and the attractiveness of new and better machines make the capitalists with expensive machines more accumulation-minded than entrepreneurs with little capital'.[1] If this was so in America in the first half of the nineteenth century, even though the resort to more capital-intensive methods did not prevent a fall in the rate of profit on capital, the effect of this fall on accumulation may have been partially offset by a rise in the proportion of profits reinvested.

We can now summarise the argument to this stage. The dearness of American labour and the inelasticity of its supply provide an adequate explanation of why, from a given range of techniques, the choice of the American manufacturer should

[1] Albert O. Hirschman and Gerald Sirkin, 'Investment Criteria and Capital-Intensity Once Again', *Quarterly Journal of Economics*, LXXII, August 1958, p. 470.

have been biased towards those which were more productive per unit of labour because they were more expensive in capital per unit of output. The same circumstances might also have exerted favourable influences on the rate of investment by providing an incentive to devise new labour-saving methods, and because capital-intensity of investment increased the ability to devise such methods and also increased the propensity to save out of profits. The main point is the favourable effect of labour-scarcity on technical progress. This might resolve the sort of dilemma emphasised by Marx.[1] It is often said that this dilemma was resolved by autonomous technical progress which sustained the rate of profit. The implication of the argument which we have been pursuing is that technological progress might itself have been the result of the exhaustion of the reserve army and not something introduced from outside the system.

The difficulty about this theory, as of any theory which regards 'restraints' as a net favourable influence on growth, is that it does not, in itself, explain why American manufacturers should have been able and prepared to continue investing in capital-intensive methods to the point where this investment yielded rapid and substantial gains in technical knowledge. Shortages frustrate as well as stimulate.

MARKET IMPERFECTIONS

Up to now we have considered the effect of the fact that American labour was dearer than the British and its supply less elastic. But we have ignored market imperfections. In the early nineteenth century, imperfections in labour- and product-markets were greater in America than in England, because population and manufacturing were dispersed over much wider distances.[2] Because American internal transport-costs were

[1] Marx himself argued in chapter 13 of *Capital* that manufacturers were induced by legislative restrictions on hours of work to look for mechanical improvements; but he did not regard the state of the labour-market as a strong impetus to technical progress. See p. 44 above.

[2] Professor Cole's data show marked regional divergencies in price movements between the Revolution and the early 1820's and a sharp decrease in divergencies in the following decades. (A. H. Cole, *Wholesale Commodity Prices in the United States, 1700–1861* (Harvard, 1938), I, p. 96.) Even as late as 1860 there was marked imperfection in the labour market. See the comparison made on 2 July 1860, by the treasurer of the Chicopee Mill in Massachusetts between this concern and the Great Falls Company in New Hampshire. 'They are at the doors and homes of a large Yankee population, and that of the best kind, while fully two-

high, the growing demand for industrial goods was met by establishing new manufacturing-centres to a very much greater extent than in England where industrial production expanded in what were already the principal industrial centres at the beginning of the century. Moreover the American manufacturer was better able than his English counterpart to exploit a given degree of imperfection; he was in less danger from new entrants into his local industry, because the size of the local market, and goodwill and other such factors hindered other people in the locality starting up manufacture.

The greater geographical dispersion of American manufacturing meant that it was less able than the English to enjoy the advantages of division of labour between industries or parts of industries.[1] On the other hand, the fact that in America problems were faced atomistically and different solutions found in different places, meant that the best solution could become widespread once improvements in communications widened the product market. We shall however not investigate such possibilities but shall concentrate on the influence of market imperfections on profits and the choice of technique.

The product-market

Because of the high cost of transport and the diffusion of industry, the *product*-market was more imperfect in America than in England. The imperfection in the product-market had a number of effects. It caused resources to be devoted to improvements in transport and in marketing and distribution. It also influenced the nature of industrial operations by favouring (*a*) the transformation of goods so bulky that they could not be easily marketed in their existing form (for example, grain) into transportable goods (for example, whisky), and (*b*) the de-

thirds of ours are Irish. But even with the Irish had we but a surplus so as to make a selection of them, I should have no fear but that we could even manufacture as cheap, or nearly so, as the Great Falls. But as it is, and has been for a year past, we have been open to a constant drain and we have been as constantly educating help to take the places of those who leave. The result is a diminished product and increased cost' (quoted in V. Shlakman, *Economic History of a Factory Town*, Smith College Studies in History, vol. XX (1935), p. 149).

[1] Market imperfections in the pre-railway age were responsible for the combining in the same establishment of the manufacture of many different types of machine. 'It is to the introduction of railroads that the advantageous subdivision of manufacture is to be chiefly ascribed. The operations of large establishments are no longer confined to particular localities.' (P.P. 1854, XXXVI, p. 109.)

centralisation of the production of goods (for example, the invention of the sewing-machine). It also shielded less efficient manufacturers from the competition of the more efficient—competition which stimulated mechanisation. These influences do not all work in the same direction. But the main point is that market imperfections enabled many American manufacturers to enjoy high profits while they confined themselves to the local markets, but made attempts to expand beyond unprofitable.

The labour-market

The imperfection of the *labour*-market within the U.S.A. was partly the result of the imperfections in the product-market which caused a large number of the additions to industrial capacity to be built in areas of new settlement where the labour-market was imperfect. It is logically distinct from labour-scarcity; for there have been countries where labour was scarce, but where the price of labour was reasonably uniform.

An American concern in the early stages of its growth could draw on the labour of the wives and children of the local farming community. There was a floor set to their wages by the need to overcome inertia and distance and, in the case of factory employment, by the need to 'make the work respectable enough to overcome the moral scruples of the community', and by the fact that they came from farming families which were more prosperous than those of English agricultural labourers. But many of these women and children cannot be said to have had a direct alternative employment in agriculture. Their alternatives were domestic service, limited in extent and not very well paid, or household work on the farms.[1] In parts of the North East this source was reinforced from the 1830's by the decline of the local agriculture which 'tended to produce a regional surplus of native-born labour'.[2] Since in the early days American manufacturing was scattered over the countryside, it must often have happened that there were few manufacturers in any given locality and that, when their concerns

[1] Caroline F. Ware, *op. cit.* pp. 11, 306.

[2] P. W. Bidwell, 'Population Growth in Southern New England, 1810–1860', *Quarterly Publications of the American Statistical Association*, N.S. xv, pp. 813–39; 'The Agricultural Revolution in New England', *American Historical Review*, xxv (1921), pp. 694–5.

were still small, they could get all the labour they wanted at a going wage set by the above factors.

But the cost of *permanent* labour, that is, of inducing whole families to commit themselves permanently to industrial work in the mill or factory, was much higher; that is, there was a differential between the price of short-term and that of permanent labour, which was much more directly related to the opportunities of alternative employment in agriculture. James Montgomery, who made a comparison of the textile industries in the two countries, wrote of America: 'The greater part of [the women employed in factories] are farmers' daughters who go into the factories only for a short time until they make a little money, and then "clear out" as it is called; so that there is a continual changing amongst them, and in all the places I have visited, they are generally scarce; on that account the manufacturers are under the necessity of paying high wages, as an inducement for girls to prefer working in the factories to housework; and while this state of thing continues it is not to be expected that wages in this country (that is, the U.S.A.) will be so low as in Great Britain. . . . The manufacturing-population of America are an entirely different class, and placed in very different circumstances from those of Great Britain, and very great changes must take place before the wages in the former can be so low as in the latter country'.[1]

Finally, once the local supplies of labour, whether permanent or short-term, had been exhausted, the real cost of obtaining additional labour might be considerable. In the New England cotton-textile industry about 1840, 'the expansion of the cotton manufacturers' industry, and the resultant demand for labour, outgrew the local resources and it was necessary for mill-owners to seek labour not only in other states or this country but in Canada and also draw upon the races which were at the time immigrating to the U.S.'.[2]

In industries where the typical concern was widely scattered —for example the woodworking industries—the supply of labour might, beyond a certain point, be very inelastic even for the individual concern. Where an industry was concentrated regionally, one manufacturer could get additional labour from

[1] Montgomery, *op. cit.* pp. 135–8.

[2] Immigrants in Industry (Immigration Commission; Washington 1910), pp. 29–30.

another, but the supply of labour was inelastic for the industry as a whole. Moreover, in the most conspicuous example of regional concentration, in the Massachusetts sector of the cotton-textile industry, the units were so large and few that the supply of labour might be inelastic to the individual concern, even though the industry was regionally concentrated. In England there was much less dispersion of industry, partly because of the different sort of power used, and in the Lancashire textile industry the firms were much smaller and more numerous than in the Massachusetts industry.

The supply curve facing many American manufacturers in the early decades of the nineteenth century was thus of a curious kind: for low outputs the marginal cost of labour was constant and might be quite low, but at some point it rose very sharply till it reached the level at which there was again something like a going wage. The marginal cost of labour must have varied a good deal from area to area and period to period and it would be worth identifying these variations in more detail. But the point to be made now is that not only was the average price of labour higher in the U.S.A., and the supply in general inelastic, but that because of imperfections in the labour-market the excess of marginal cost over average price was for many American manufacturers much greater than in England.

The imperfections of the labour-market did not uniformly favour the adoption of machinery. Even a sharp rise in marginal labour-costs, as local supplies were exhausted, may not invariably have been favourable to mechanisation. It is not necessarily the case that rising marginal labour-costs are a stronger stimulus to the adoption of mechanical methods when the rise is abrupt than when it is gradual. A moderate general shortage of labour may be more stimulating than a severe local shortage. For with a reasonably perfect labour-market and gradually rising marginal labour-costs, those firms which do not attempt to compensate by installing labour-saving machinery will be threatened by the competition of firms which do. Whereas, in a highly imperfect labour-market where labour is unobtainable beyond a certain point, a firm which does not mechanise may suffer no disadvantage beyond the fact that it does not expand; and it may very well decide that expansion is not worth the trouble since it involves considerable reorganisation. All that the imperfection of the labour-market by itself implies is that, for relatively small outputs, average profits were high and that, if a firm wanted to expand beyond this, it might *have* to mechanise.

Imperfection in the labour-market meant that, up to a certain point, the American manufacturer could exploit his labour; imperfection in the product-market meant that up to a certain point he could exploit his customers. But beyond this point he could sell additional output only by considerably lowering his price, that is, with a fall in his marginal rate of profit per dollar of capital employed. Where the U.S. manufacturer was a monopsonist in his labour-market and an oligopolist in his product-market, he got higher average profits than the English manufacturer for the early stages of his output. Up to this point we could explain why U.S. investment should be faster than the British, even if we assumed that capital-markets were perfect within the U.S.A. and that capital moved freely between the U.K. and the U.S.A. With a higher rate of profit, American manufacturers were prepared to have more at risk at any given time, that is, were prepared to undertake a larger increment of new investment in relation to present wealth. Until the point where marginal returns threatened to fall came in sight, the American manufacturer would be prepared to have a higher proportion of his wealth in unfinished projects than the English, because the projects looked like being so profitable that it was worth his while to stick his neck out further. On this reasoning the imperfections in the labour- and product-markets explain why, in the early stage, the *rate* of investment might be faster in America than Britain, but they do not, at this stage, reinforce, and they possibly weaken, the bias towards capital-intensive investment given by labour-scarcity in general and the inelasticity of its supply to the industrial sector as a whole.

If, with further expansion of capacity, the fall in the marginal rate of profit was precipitated from the product side—by the exhaustion of the local market possibilities—and while the manufacturer was still enjoying the advantages of a monopsonist position in respect of his labour, he would have an incentive to search out methods which reduced his capital and labour per unit of output, but no bias in a particular direction (assuming that wages were already at the floor and assuming no imperfections in capital-markets). If the limitation appeared on the labour side while he was still enjoying an oligopolistic position in his product-markets, he would have an incentive to adopt more capital-intensive methods, but only to the extent that the rise in price necessary to cover the cost of additional labour would attract competition from other areas. When his marginal rates of profit were under pressure from both sides he would have a general incentive to cut costs, and a particular incentive to adopt the capital-intensive methods which cut labour-costs. Moreover, since his average profits were high he had the means to acquire the necessary capital-intensive equipment.

The imperfections in the product- and labour-markets are therefore capable of explaining why additions to capacity should have been biased towards capital-intensity, and how internal finance was available to undertake the additions. But no additional light has been shed on the question why, in the face of the sharply falling profits which the situation assumes, the American manufacturer should have wished to expand his capacity at all. If he were to expand, he would do best to buy capital-intensive equipment, but why in these circumstances should he want to buy new equipment of any sort? Why should he not instead devote his high average profits to other purposes? Why, in fact, did American industry ever get beyond the stage of a network of small manufacturers exploiting their position as monopsonists of labour and oligopolists of products? We still have no additional explanation to that offered in the discussion of labour-scarcity that the capital-intensive composition of the investment might have yielded great and rapid technical progress.

The capital-market

The impediments to the movement of funds within the U.S.A. were greater than in England. Whereas in England there were no significant regional variations in interest rates or in availability of funds, in the U.S.A. the differences were striking. The interest rates quoted on loans in the Ohio valley in 1827—between 10 and 20 per cent on large sums—compared with the rates in Massachusetts suggest considerable geographical immobilities; and the fact that interest rates for small sums in the Ohio valley in the same year were as high as 36 per cent implies considerable imperfection within that area itself.[1] Apart

[1] T. S. Berry, *op. cit.* p. 411.

from the normal sources of market imperfection, state law or custom imposed restrictions on long-term loans by institutions to borrowers outside the state. Extreme provincialism, says one study, appears to have marked the portfolios of savings banks in New York, Pennsylvania and Maryland.[1]

As a result of these regional differences, capital in some places was very dear by American standards; but by the same token in other places it was cheap, and in such places the general bias towards capital-intensity was reinforced. It was further reinforced if these places were also areas where because of imperfections in the labour-market, labour was dear by American standards; and though there must have been a general tendency for dear capital and dear labour to coincide, it did sometimes happen that the area of cheap capital was also an area of dear labour.

There was a second type of imperfection in the capital-market which concerned the ease with which funds could move between industry and other employments. This has to be considered from two aspects: the ease with which the manufacturer could borrow from other people when he wished to expand his capacity faster than the ploughing back of his profits enabled him to do; and the ease with which the manufacturer who did not wish to plough back all his saved profits could find an alternative use for them. The more difficult it was for a manufacturer to borrow to extend his operations, the weaker the bias towards capital-intensity provided by a given level of labour-costs. By contrast, when a manufacturer wished to place his savings outside his concern, the more difficult it was to find outlets, the lower the opportunity-cost of his capital and therefore the greater the bias towards capital-intensity.

There is a general presumption that access to outside funds was easier in England than in America, since interest rates were lower, and after 1815 there was a large volume of *rentier* savings seeking a home. But it is by no means certain that this presumption is correct. For many years during the wars against France it was extremely difficult in England to borrow money on a mortgage—the instrument most commonly used by a manufacturer—for the government was borrowing heavily, and

[1] L. E. Davis, 'The New England Textile Mills and the Capital Markets: A Study of Industrial Borrowing 1840–1860', *Journal of Economic History*, xx (March 1960), p. 4.

the rate on mortgages was subject to an effective legal maximum below the rate obtainable on government loans. While there were also usury laws in America they do not seem to have operated in quite the same way, and it is possible that, when the war of 1812 curtailed the possibilities in commerce, capital was easier to get for industry in Massachusetts than in England. After 1815 though funds were abundant in England, English *rentier* investors were interested, not in industry, but in government bonds; and the savings which before 1815 went into British government debt, after 1815 went into the bonds of foreign governments and into housing and transport improvements. In America, funds diverted from government debt were more likely to go into industry. 'Between 1822 and 1826' write Professor and Mrs Handlin, 'men like Henry Lee and Harrison Gray Otis, without other outlets for investment, began to put into manufacturing corporations large sums released by a contracting public debt'.[1] It is true that the abundance of funds in England provided short-term finance for industry and so made it easier for industry to pay for additions to plant from its own resources. But English funds also went to a considerable extent to finance foreign trade as well. Moreover, the fact that the privileges of incorporation were much more easily acquired in America than in England in the first half of the century gave certain types of American industry readier access to non-industrial savings, and indeed, it was sometimes said, gave them readier access even to English 'blind capital' than English industrialists themselves enjoyed.[2] When, in addition, the impediments to the free movement of capital within America are taken into account, there may have been places in the U.S.A. where access to capital from sources outside the firm was easier than it was in England.

In neither country, however, is it likely that borrowing was a major source of industrial capital; extensions of capacity were mainly financed by the ploughing back of profits. It is probable, therefore, that manufacturers appeared on the capital-market more often as lenders than as borrowers, and that, so far as its

[1] O. and M. F. Handlin, *Commonwealth, a study of the role of government in the American Economy; Massachusetts, 1774–1861* (New York, 1947), p. 197.

[2] For English investment in American securities at the end of the eighteenth century see J. S. Davis, *Essays in the Earlier History of American Corporations* (Cambridge, Mass. 1917), I, pp. 144, 149, 152–3, 186, 193, 282, 307, 360, 364–5, 447; II, pp. 39–40, 169, 279, 299.

influence on industrial finance was concerned, the capital-market was less important as a source of funds than as an alternative use for industrial profits.

It is obvious that there was more to be done with capital in the U.S.A. than in England. This, however, does not settle the question of the range of alternatives available to manufacturers. In America the manufacturer could purchase land, particularly for speculative purposes. But the speculative pull of land in America has to be matched against its social pull in England. The purchase of an estate yielded a psychic return in England, where landed estates were a source of political power, which they did not yield in America, where indeed the accessibility of land and scarcity of labour made it difficult to sustain an estate form of organisation. 'In general in America', wrote de Tocqueville, 'one cannot find tenants. . . . Land costs too little and its products are too cheap for anyone to want to cultivate it unless he is the owner. Without tenants no great territorial fortunes'. And again, 'America does not at all favour the existence of great landed fortunes, and the landowners were never able to get large incomes from their lands.'[1] Closely associated with the acquisition of a landed estate was the establishment of a social position. The English manufacturer was much more likely than the American to spend money on his house, on the education of his children, on supporting a style of life which was ultimately set by the aristocracy and gentry.

Government securities, a second alternative, were probably not generally attractive to manufacturers in either country, for their return was in almost all circumstances likely to be lower than that on industrial investment; the market in bricks and mortar probably did not much overlap with the market in government paper. But in so far as they did attract industrial funds they were more likely to do so in England. The market in British government securities was long established and well organised. Because England had a large class of *rentier* investors, a large volume of foreign government debt was also available in London; the market arrangements of American state debt were biased towards England. Furthermore, the existence of a large national debt made it easier for the English manufacturer who was anxious to curtail his stake in industry

[1] de Tocqueville, *op. cit.* pp. 49, 73.

and become a landowner to give effect to his ambitions; for it enabled him to hold a substantial part of what remained of his wealth after he had bought landed property in a form which, while it yielded less than an industrial investment, yielded more than land and required no management. Richard Arkwright, Jnr and Sir Robert Peel, Snr both had substantial holdings in the Funds in 1812, when the marginal yield on industrial investment must have been high; and the Welsh ironmaster, William Crawshay, was reported to be at his death the largest single holder of Consols.[1]

Investment in transport improvements was an attractive alternative in both countries, particularly to those manufacturers who hoped to gain either by contracting for materials or from the fall in transport-costs. Industrial money in both countries went into rails. Another possibility was the mortgage. This played an essential role in the expansion of the American frontier; on the other hand there was in America no credit-worthy but unthrifty aristocracy and gentry of the kind which in England borrowed heavily on mortgage, and land and materials for house building were cheaper, so that the urban house-mortgage played a less important role in America.

Without more information than is readily available about the ways in which manufacturers in the two countries disposed of their resources one cannot generalise, but the possibility cannot be ruled out that the British manufacturer had more to do with his money outside industry than the American, and that, at a given marginal rate of profit, he was less disposed to reinvest his profits in his own concern. Whether or not this was true in general, it must have been the case that there were some areas in America where the alternatives available to a manufacturer were more restricted than they were in England. For in England there was virtually no part of the country in which a man could not easily find a home for his savings through a local solicitor or a bank, and, in the case of mortgages and government debt, an investment with very few risks; whereas in the U.S.A. the various capital-markets were young and poorly organised, many of the assets available involved a high degree of risk, and lenders and borrowers were unevenly distributed over the country. 'Small enterprises' writes a historian of American industry 'catering to local markets and

[1] J. P. Addis, *The Crawshay Dynasty* (Cardiff, 1957), p. 157.

managed directly by their proprietors, probably afforded as safe employment for money as any other sphere of production'.[1] They were risky, but so was everything else. Where for such reasons the American manufacturer had negligible attractive alternatives to reinvestment, the opportunity-costs of industrial capital were lower than might be supposed from the overall scarcity of American capital, and might on occasion conceivably be lower than in England.[2]

In so far as the English labour- and product-markets were more perfect than the American, the English manufacturer could continue to get his average rate of profit on his investment simply by duplicating capacity of the existing type; even if his average rate of profit was as high as the American, he was not under pressure from falling marginal rates to search for methods which mitigated the fall. But *even if* the English manufacturer were faced with an analogous fall his response might be different. If he were a very substantial manufacturer, for example, the possibility of obtaining a position in English society set a limit to the fall in marginal profit-rates which he was prepared to suffer and still devote his profits to the expansion of business.

Thus the imperfections in the labour- and product-markets gave the high average rate of profit which provided the American manufacturer with the means to buy labour-saving equipment; they also explain why, beyond a certain point in the expansion of output, he needed to adopt such equipment if the low marginal rate on capital was to be alleviated. Wherever, in addition, attractive alternative uses for his profits were not easily available, the manufacturer had a reason why he should continue to attempt to expand his capacity even when the marginal rate of profit on capital was falling. Moreover the absence of alternative uses for his funds not only provided a reason why he should persist in investment; it reinforced the bias towards capital-intensive techniques. Given that he was going to reinvest in any case, an American manufacturer in the situation envisaged would be best served by adopting those techniques which yielded him the largest total revenue after

[1] Clark, *op. cit.* I, 370–1.

[2] Imperfections in the capital market appear to have been favourable to Massachusetts. 'In 1831 Massachusetts money sought manufacturing investments freely on the prospect of 6 per cent returns' (Clark, *op. cit.* I, p. 369) and in the 1840's, in times of normal activity, the free market equilibrium rate of interest seems to have been not far from the legal maximum of 6 per cent. (Davis, *op. cit.* p. 9.)

payment for his capital. Where increase in output would lead to a fall in product prices, because his product-market was imperfect, he would be more interested in reducing costs per unit, or in improving facilities for marketing than in increasing output; he would be more interested in a technique which employed all his capital rather than one which employed a part of it at a higher (but not proportionately higher) rate of return per unit of capital. And where his marginal labour-costs were threatening to rise he would be interested in the technique which did most to mitigate the threat. That is, provided he could get a higher return than he could get elsewhere, he would prefer the method which, because it created least disturbance in his labour- and product-markets, gave him the best return on *all* the capital he had to invest, and he had a strong incentive to search out such a method.

Thus the imperfections in the capital-market explains why in some circumstances American manufacturers should reinvest their profits and also why the investment should be capital-intensive. These effects reinforce each other. The manufacturer reinvests his profits because the alternative uses are restricted; but this reinvestment increases the probability that, unless his investment is very capital-intensive, his marginal labour-costs will rise sharply. And the effects are cumulative. If technical progress yields a high profit to a firm, its rate of investment will increase and the bias towards capital-intensity will be strengthened.

The purpose of this exercise has been to examine the various ways in which market imperfections might have influenced the choice of techniques in America; the object has not been to measure their practical importance, which clearly changed considerably as a result of the public improvements of the period. The general effects of market imperfections did not uniformly work in the same direction, but they did sometimes result in situations which strengthened the bias towards the adoption of labour-saving capital-intensive methods provided by the general character of America's factor-endowments. Where the capital-intensive methods adopted in such situations led to technical progress they might then exert an influence in other areas where the effect of market imperfections was less favourable to the introduction of labour-saving techniques.

We shall now summarise the argument in a slightly more

formal way with the aid of rather extreme assumptions made for the purposes of exposition rather than for their exact correspondence with the facts. There are two logically distinct parts to the story, the first relating to the scarcity and inelasticity of labour-supplies in the U.S.A., and the second to various types of market imperfection.

It is possible with a small number of assumptions to explain why dearness of labour should have given American investment a capital-intensive bias. Suppose that labour was 30 per cent dearer in the U.S.A. than in England, that is, that there was a reservoir of agricultural labour in both countries from which industry could draw additional labour at a going wage, but the going wage was 30 per cent higher in the U.S.A. because land was plentiful and productive, and because labour, while moving freely within the U.S.A. and within England, did not move between them sufficiently to remove the disparity. Suppose also that product-prices in the U.S.A. were higher than in England by sufficient to ensure the same level of profits in both countries, that is, that the tariff was high enough to offset the net effect on profits of dear American labour. If in this situation the prices of capital goods and interest rates were the same in both countries and unchanging, the Americans would adopt more capital-intensive techniques than the English. So long as additional labour was available to American industrialists at the going wage the *preference* for the capital-intensive techniques would not increase, but if autonomous technical progress produced a flow of new techniques the Americans would make a more capital-intensive selection than the British and the composition of investment in the two countries would increasingly diverge.

If American labour were less elastic as well as dearer, the bias towards capital-intensity would increase; for the increase in labour-costs would raise the cost of the labour-intensive methods while the price of imported machines remained the same. Even if the Americans had now to pay more for their machines than the British, the inelasticity of supply of American labour would still give their investment a capital-intensive bias (though in this case a weaker one) so long as the tariff was not raised to compensate them for the rising cost of labour. We can, therefore, on these lines explain the composition of investment.

But, on these assumptions, we cannot say much about the

rate of investment. If we persist in the assumption of perfect capital-markets within the U.K. and the U.S.A. then the rate of investment in each, and the disparity between them, are indeterminate. If we assume that the rate of investment was related to profits in the same way in both countries, it would be subject to a limitation in the U.S.A. to which it was not subject in the U.K. The rate of profit in America would not fall, so long as American industrial investment did not expand faster than agriculture was prepared to release labour and than population increased. But given the inelasticity of American labour, an identical rate of investment in both countries would run up against rising marginal labour-costs earlier in the U.S.A. than in the U.K., and, on the assumption that the tariff was not adjusted to offset them, against falling rates of profit. The same facts that explain the *composition* of American capital make it more difficult to explain its *rate*.

It is possible to resolve this dilemma if we suppose that the composition of investment had a marked effect on its rate by (*a*) making manufacturers more machine-minded and by enabling them to make technical progress and (*b*) giving American manufacturers a high propensity to save. But it is not necessary to rest the whole argument on these suppositions. Some reasonable assumptions about market imperfections will reinforce them.

We retain the assumptions of a competitive capital-market in both countries, the same price for capital goods in both countries, the same rate of interest, and the cost of labour 30 per cent higher in America, but the labour-market perfect though the supply inelastic. If the tariff has been fixed at a level *more* than sufficient to compensate for the higher cost of American labour, imperfection in the product-market would prevent American manufacturers undercutting each other to the point where the tariff is no more than sufficient for the purpose. Where, because of the dispersion of industry and the difficulties of internal transport, each local market was supplied by one local manufacturer and by imports, the local manufacturer would be able to charge a price equivalent to the c.i.f. cost of import plus the tariff. In this situation the scarcity and inelasticity of labour would give the bias towards capital-intensity and the imperfection of the product-market would give high average profits.

Imperfection in the labour-market also helps to explain rate and composition—the high average profits give the means, the low marginal profits give the bias towards capital-intensity (this would be so even if the product-market were not imperfect).

Imperfection in the capital-market has divergent effects. If the industrial capital-market was more imperfect in the U.S.A. than in England, American industry would have had to pay more for its capital and so would have had less bias to capital-intensity than it would have had in the absence of such imperfection. If we assume that within America the market in industrial capital was more imperfect than the product- and labour-markets, this further tells against capital-intensity. On the other hand if the imperfections of the capital-market in America took the form of limiting the range of opportunities available to the manufacturer outside his own concern, this told in favour of more capital-intensive methods. Taken in conjunction with the imperfections in the labour- and product-markets which produced high average profits, this second type of imperfection in the capital-market would help to explain both rate and composition; the rate because it provides a reason why manufacturers should have wanted to invest even at low marginal rates of profit on capital, the composition because it gave manufacturers a bias towards techniques which made full use of their funds.

This argument has been conducted in terms of the effect of labour-costs on the marginal rate of profit on capital, that is on the assumption that manufacturers were primarily interested in maximising their internal rate of profit per dollar of capital employed. It should be observed here that all these arguments would be stronger if the marginal rate of profit in America fell below the British rate. But it is not necessary to the argument to assume that this happened. The threat or expectation of its happening may have been enough to stimulate successful attempts to prevent its ever happening. It may also be that what was important for the choice of techniques was not the absolute levels of profit-rates in the two countries but the rate at which these rates threatened to fall for a given amount of additional investment, that is, a fall from, say, 12 per cent to 10 per cent may have provided American entrepreneurs with a stimulus to move towards the capital-intensive end of the spectrum which the English, at a constant, say, 10 per cent, lacked. Americans may have become habituated to a higher rate of profit than the British and more sensitive to any divergence from it—more sensitive, that is, in their choice of technique but not in their desire to invest.

Moreover, the essence of the story would still be preserved even if one supposed that the manufacturers were not primarily interested in maximising their internal rate of return on their capital. It has been assumed that this was the principal aim of manufacturers, since this probably is the assumption which best fits the facts of the early nineteenth century. If, in fact, business men were more interested in the preservation of the business entity or in maintaining or improving the market position of their firm as measured by sales, labour-scarcity would show itself in the difficulty of attracting labour at some desirable rate of growth or even difficulty in retaining labour, and this difficulty would provide an incentive to move to techniques which had high productivity per head because they were expensive in capital. It may very well be the case that the American manufacturer was more concerned with output per worker than with the marginal rate of profit; that is, provided a project would yield some minimum rate of profit he was more interested in its effect on productivity than on profits and might opt for a method which, because of its high labour productivity or because it improved conditions of work, enabled him to retain his labour, in preference to one which—in the short run at least—yielded a higher rate of profit.

LABOUR-SCARCITY AND THE TRADE CYCLE[1]

We shall now attempt to take account of changes in the supply of labour through time and particularly over the course of the trade cycle. In the context of our argument, a trade-cycle boom can be looked on in two lights: (*a*) as a period when investment was high and therefore the opportunities greatest for introducing new methods of any kind and (*b*) as a period when labour-shortage was particularly acute, and therefore the incentive strongest to undertake investment of a particular, that is, capital-intensive, kind.

High investment

The contrast between England and America in the first half of

[1] On this section see N. Kaldor, *Essays on Economic Stability and Growth* (1960) 10 and 13.

the nineteenth century corresponds to some extent to the contrast in theory between an economy in which profits are a residue after the payment of subsistence wages and one in which they are determined by the rate of investment. In America there were several circumstances which favoured a very high rate of investment during a boom: the strong mania element injected by land-speculation which inflated the expectations of business men; the importance in the economy of transport and building which are always heavy absorbers of capital. In these circumstances, it might be argued, American manufacturers might have been able to wrest from consumers and labour sufficient resources to maintain a higher rate of investment than their English contemporaries. In so far as they were successful in doing this, that is in so far as by investing they raised the rate of profit, this would reduce their incentive to introduce capital-intensive methods; but, despite this, if the long-term scarcity of labour had ensured that a high proportion of the methods awaiting employment were capital-intensive, investment during the boom would take a predominantly capital-intensive character. This is another way in which the capital-intensive bias given to the *composition* of American investment by labour-scarcity might conceivably be reconciled with a favourable effect on the *rate*. Is this sequence of events likely to have been important in practice?

It is reasonable to suppose that the optimism of American entrepreneurs was more easily roused than that of English entrepreneurs, and that American investment was more responsive than English to a given increase in profit expectations. But it is also true that, in certain types of investment, a higher degree of optimism was necessary in the U.S.A. in view of the job to be done. 'To make a railway where population exists is one thing', said a nineteenth-century railway-builder, 'but to make one in order to call forth a population is another'.[1] In America the minimum feasible size of railway system was very large in relation to existing resources—so large that only people who felt highly optimistic would undertake railway-building. In the case of America therefore there is force in the argument that the cycle determined the trend, and that the influences which engendered extravagant hopes during the

[1] Watkin, *op. cit.* p. 125.

boom enabled investment to take place which otherwise would not have been undertaken at all. In England on the other hand the minimum size of viable railway system was small, and the English railway system would have been built in any case, and with less waste, if it had been built gradually rather than, as it was, concentrated in booms.

But, for our present purpose, the important point is how the vigour of American booms affected American manufacturing industry. And, in this field, circumstances imposed severe limitations upon the ability of American manufacturers, by stepping up investment, to wrest resources from consumers and labour. The ability to wrest resources from the former was restricted by imports which limited the rise in American product prices in the case of internationally traded goods. The ability to wrest resources from the latter was set by the inelasticity of American labour-supplies. It was much more difficult for American than for English manufacturers to get their own way with labour in a boom; and the position of labour during a boom was further strengthened by the fact that the American booms usually involved transport-investment and the opening up of new areas which exerted a pull on the labour of the regions of earlier settlement. American industrial investment during the boom was curbed by the rising cost of labour, directly through its effects on profits, and indirectly through the effect of the balance of trade—the heavy demands for imports, not covered entirely by foreign borrowing, leading to pressure on the exchanges and curtailing the supply of liquid assets.

If one can judge from contemporary comment, during the boom, labour was the limitation in America as it was not in England.[1] And, moreover, the rise in American wages might not measure the full effect of labour-shortage on the investment policy of business men, since if they thought that investment

[1] C. Ware gives evidence of labour-scarcity in most periods of high activity: in 1793 when mills 'were being established further back into the country in order more easily to secure workers'; during the expansion of the war years; in 1821 and 1831 and in the boom of 1846. 'The multitude of mills which the war prices of goods have recently caused to be erected ... employs all that can be found', wrote one manufacturer in 1814. In 1821 the agent of one company wrote of 'the standing still of the looms for want of weavers'. 'Hands are becoming scarce', wrote an agent in 1846 (*op. cit.* pp. 201, 214, 227–8, 304.) For labour-shortage in the revival of activity after the 1857 crisis, due to the competitive demands or other industrial centres, see Shlakman, *op. cit.* 'This company found that its labour costs were increasing on account of a scarcity of good workers, and the lack of a

was proceeding faster than, from past experience, labour was likely to be forthcoming, they might draw in their horns in advance of any sharp increase in wages which would therefore not occur.

Labour-shortage

It is, therefore, more reasonable to look on the boom, not as a period when the American rate of profit on industrial investment rose more than the English, but as a period when labour-shortage became most acute, and when European products tended to enter the U.S.A. so that the ability of the American manufacturer to pass on his rising labour-costs was most restricted. During a boom, moreover, demand curves would move sharply to the right, and thus in conditions of market imperfection the manufacturer was likely to run up against labour-scarcity *before* his market was saturated (see the argument on p. 69). In products where the ceiling to American prices was set by the entry of European goods over the barrier imposed by the tariff and transport-costs, the demand-curve facing the individual American producer would become much more price-elastic: the inelasticity of demand only occurred because of his sheltered market. There would therefore have been a very strong inducement, in any investment undertaken during a boom, to shift to more capital-intensive techniques—an incentive, where the supply of labour was extremely inelastic, to jump over a gap in the spectrum of techniques.

The labour-scarcity of the boom would be followed by a shift to the right of the short-term supply curve of labour as migration, internal and international, started; and this movement would be accentuated by the normal growth of population, the success of labour-saving innovations made during the boom, and the fall in output during the slump. During the recession, labour, though scarcer than before the last boom, was relatively plentiful again. The best way to cut costs—and

surplus from which to choose the best' (pp. 147–9). 'A change in the form of models or parts affords a favourable opportunity for the operatives to press an increase of their wages. ... There are likewise periods in the general business of the country when labour and provisions advance in price. Such times are always seized on to increase their wages. When a revulsion takes place and prices elsewhere sink to their former level, it is no easy matter to reduce the wages of armourers. We have witnessed this state of things several times during the last 25 years.' (An Ordnance Officer on Springfield in 1841, quoted in Deyrup, *op. cit.* p. 173.)

as demand fell this is what manufacturers attempted to do—was, therefore, to cut overheads by increasing the running hours of machinery, shifting to more labour-intensity with a fixed amount of capital, and improving the performance of the capital-intensive methods introduced during the boom. If the possibilities of such improvement were considerable, average profits over all phases of the cycle might be higher than in England, while the rate of profit for a given expansion in the boom might be low. Again, the absorption of labour by firms which, during the depression, shifted to more labour-intensity with a fixed amount of capital, would to some extent offset the fall in demand for labour by firms which closed down and the increase in the supply of labour due to population-growth, etc. Thus conditions would be created in which the next boom, when it came, might again encounter an inelastic labour-supply.

This mechanism would obviously not work when the slump was very severe, for the situation envisaged means that some firms were actually increasing their output during the depression. It would work most effectively when many manufacturers went out of production in the early days of a slump, thus allowing the remainder to increase sales to a falling or static market and accumulate large enough profits for the next boom. On the whole, the fall in total output in the slumps of the first half of the century seems to have been concentrated mainly in investment goods, and it is with consumer goods that this argument is concerned.

EXPANSION OF THE MARKET

In most of the preceding discussion we have assumed that the rate of industrial investment was limited by the supply of factors and depended on the rate of profit. We have assumed that, via the rate of profit, the level of investment in both the U.S.A. and in England was adjusted to the rate of increase of supply of all factors, and we have considered the consequence of the fact that, when investment increased at these rates, shortages of one particular factor, labour, appeared earlier in the U.S.A. than in England. But the rate of investment also depended on the expansion of the market. We must now consider the possibility that, in relation to the total supply of factors to industry, the demand for American manufactures in the first half of the nineteenth century was expanding more rapidly than the English.[1]

This is prima facie probable. The major part of the demand for American manufactures came from the rise in agricultural incomes as cotton exports from the southern states rose and as the country was opened up and population increased.[2] In contrast, though a substantial part of the demand for English manufactures also came from an increase in consumers' incomes in primary producing-areas, a considerable part came from a switch of demand from domestic-type industry to English factory industry, and depended on a fall in the cost of the English products. Thus the long-term growth in the demand for American industrial goods probably warranted a rate of investment, in relation to the supply of factors as a whole, which was more rapid than that in England.

Moreover, it is probable that any given increase in demand was likely to lead to a larger increase in investment in America than in England. In England an increase in demand was met from the existing centres of production where there were generally some possibilities of increasing output with small changes in existing equipment. In the U.S.A., because of the imperfections in the product-market which we have already discussed, an increase in demand in a new area of settlement was likely to be met by the creation of new capacity and new concerns within the area, even though there may have existed some slack in the older area, and without corresponding disinvestment in the older areas.

The rate of investment in relation to total factor-supplies is relevant to our argument in two ways. It is relevant, first, to the effect of factor-supplies on the invention and adoption of

[1] This presumably is what contemporary observers like Gallatin (*op. cit.* p. 430) meant who said that American development was retarded by both dear capital and dear labour.

[2] In the early stages of industrialisation the expansion of exports of primary products from southern states was of critical importance. It is unlikely that these export staples could have been profitably produced by dear white labour. 'In a country where labour is as expensive as it is in America' wrote de Tocqueville, 'it would be difficult to grow tobacco without slaves' (*op. cit.* p. 77). Thus in a sense the industry of the northern states was able to escape the consequences of dear labour by mechanisation because the southern states, where crops were labour-intensive and mechanisation would have been technically difficult, had been able to escape the consequences of dear labour by importing it.

new methods. To the extent that investment was pressing more closely on total factor-supply in America than in England, the advance of technology would have been more sensitive to factor-endowment in America. Technology may be expected to edge along, adjusting itself to relative resource-scarcities, only when there is, to begin with, a rough balance between resources and investment—when the increase in capacity is constantly pressing on the available supplies of labour, natural resources and finance. It is in these circumstances that manufacturers are most likely to get clear indications of relative resource-scarcities, that those who adopt methods which are inappropriate to the relative resource-scarcities of the economy are penalised and that a new technique—whatever resource it saves—is likely, when introduced into one firm or industry, to have repercussions on other firms or industries and force them either to contract or innovate. Thus the firms in the U.S.A. which in the 1820's and '30's adopted labour-intensive techniques would have been placed in a disadvantageous position in competition with firms in the same or competing industries which had made a more appropriate choice and had adopted capital-intensive techniques. But this was not true to the same extent of firms which adopted capital-intensive techniques in England, the main effect of which would simply have been to depress wages. For the same reason new techniques which increased the productivity of all inputs had a more general effect in the U.S.A. than in England. If, in the early decades of the century, the actual rate of growth was higher in the U.S.A. compared with the possible rate of growth, this fact must be counted as a circumstance favourable to technical improvement, quite apart from the relative scarcity of labour *vis-à-vis* other factors.[1]

The fact that investment in America was rapid in relation to the supply of all factors except agricultural land gave an incentive to improvements which were capital-saving as well as to those which saved labour. Where the existing techniques afforded no possibilities of alleviating the dearness of labour but the entrepreneur still persisted in the attempt to widen,

[1] Though heavy demand might sometimes blunt the edge of the incentive to improve technique, 'The cotton-textile machinery industry of the 1870's had been softened by three or more decades of heavy demand . . . in which the advance in machine technology had been very slow'. (Navin, *op. cit.* p. 109.)

the difficulty of getting capital would induce him to concentrate on making the most of his machinery in all ways which did not involve the use of additional labour per unit of output. Where the range of existing methods *did* allow the entrepreneur to compensate for dear labour by substituting capital, the adoption of more capital-intensive techniques would provoke the problem of finance even more acutely, especially if the techniques were very effective in saving labour. When investment is pressing against resources as a whole, the temporary resolution of the most severe scarcity is likely to be followed by the emergence of scarcity in some other factor.

A 'scarcity' of capital not only provided the user of machines with an inducement to get the most out of them: it gave the manufacturer of equipment an inducement to provide them as cheaply as possible. Many of the goods, the manufacture of which was most highly mechanised, were not single-use consumer goods but equipment designed to increase labour productivity or at least to meet a production problem (like the steel ploughs which were necessary to open up prairie soils). Where the cost of the minimum feasible piece of such equipment was large in relation to the funds available to the typical user, the demand was very sensitive to its price. This was said to be the case with woodworking-machinery,[1] and it was probably also true of some types of agricultural machinery, of such goods as sewing-machines and of such activities as shipbuilding. This was one reason for the flimsiness of much American equipment. According to the American friend of de Tocqueville whom we have already quoted, 'one reason why our ships do not last long is that our merchants often have little capital at their disposal to begin with. It is a matter of calculation on their part. Provided that the ship lasts long enough to bring them in a certain sum beyond their expenses, their aim is attained'.[2]

Almost every observer pointed to this contrast between the durable English and the flimsy short-lived American equipment. For example, writing of woodworking-machines, an English author observed 'For builder's machines to supply the

[1] Richards, *Treatise*, *op. cit.*, pp. 35, 50–51. 'The (woodworking) machines are for the most part sold to men of limited means, who have not only to consider the worth of the money after investment, but have first the greater difficulty of commanding a sum sufficient to purchase the machines.'

[2] de Tocqueville, *op. cit.* p. 111.

American market they must be cheap, capable of doing the maximum amount of work when operated with first-class skill. . . . For the English home market machines must be better fitted to correspond to the general character of other engineering work. Changes not being so frequent, and the first cost less, they can be made heavier and stronger'.[1] For this general contrast there are many reasons. The flimsiness of American construction may have been partly the result of the technical inability of American engineers to make high-quality durable machines, and to the high cost of iron and the low cost of wood. It was partly the result of the fact that American engineers were not so long-established a profession, applied less rigorous technical standards and so allowed more weight to be given to economic considerations, in contrast to English engineers who were apt to subordinate economic considerations and who sometimes boasted of the fact. 'We would not be understood', wrote the author we have just quoted, 'as making the plans and designs of the engineer subservient to commercial conditions'.[2] If British purchasers of machines had consulted exclusively economic interests, unconfined by the technical prejudices of engineers, they might have preferred cheaper and less durable machines. But in addition there were quite rational economic reasons why Americans should have attached more importance than the English to building cheap machines of the sort which would bring quick returns. There was the need discussed earlier to modify the burden of the large amount of capital involved in the choice of capital-intensive techniques; though the American chose the more capital-intensive of existing techniques, any given technique was apt to be embodied in a less durable form in America than in England. This need made the American purchaser of machines readier to accept standard products, and less inclined to force upon machine-makers minor, but expensive, modifications. Then there was the expectation of more rapid technical obsolescence. But all these reasons had greater force in so far as investment in America was pressing closely on capital as well as on labour.

In building transport-systems, too, the Americans attempted to economise capital more resolutely than did the English. Edward Watkin, a leading English railway-builder, put the

[1] Richards, *Treatise*, p. 53. [2] *ibid.* p. 51.

contrast between the methods of the two countries at the middle of the century in the following terms: 'The cost of American lines has been brought down by the necessity of making a little capital go a long way, and by the sacrifice of many of the elements of permanent endurance which attach to our railways. We have deemed the inventions of railways a final improvement in the means of locomotion, and we have, therefore, constructed our works to last "for ever" of bricks and mortar. We have made our rails strong enough for any possible weight of engine; our drainage capacious enough to remove any conceivable flood ... our bridges firm enough for many times the weight that can ever come upon them'. The Americans believed that the English desire for permanence was 'a bar to future improvement; while their [the American] plan for putting up with "what will do" leaves the door open for invention.'[1] These were typical of the views held by most people who had experience of the two countries. 'In making railways in the United States', said an American witness before the Committee on the Export of Machinery, 'we aim to economise capital, and we therefore are not so particular about reducing the gradients as you are'.[2] 'In the construction of railways', said another observer, 'economy and speedy completion are the points which have been specially considered. It is the general opinion that it is better to extend the system of railways as far as possible at once, and be satisfied in the first instance with that quality of construction which present circumstances admit of, rather than to postpone the execution of work so immediately beneficial to the country ...'. Whitworth quotes the case of the railway connecting the east and west parts of Pennsylvania, which in 1854 was carried over the Allegheny Mountains by a series of five inclined planes, operated by stationary engines. A new continuous road was then under construction which would dispense with the inclined planes and save four hours. 'It is doubtful whether the delay would not have been very considerable, had the construction of the railroad been postponed until means had been found for executing these great works in the first instance'.[3] The American canals seem also to have been built quickly and cheaply. 'They have built the longest canal in the world in the least

[1] Watkins, *op. cit.*, p. 124. [2] P.P. 1841, VII, Qu. 3012.
[3] P.P. 1854, XXXVI, p. 127.

time for the least money' wrote one observer about the Erie canal.[1]

At least in the case of railroads, however, there are some explanations of flimsy and rapid American construction which are independent of desire to economise on capital. Thus, over its entire life-time, an American railway was likely to be less intensively used than an English railway and therefore did not need to be so well built. There was also in America a much greater disparity between intensity of usage in the immediate future and in the longer run. In America a system big enough to carry the load expected when the new region was fully opened up would be much too big for the traffic in the years immediately after building; and therefore the railways were built in ways which allowed them to be modified most easily when the increased demand came.[2] The English, who could expect the load soon after building to be not dramatically below its maximum, had no reason not to build to last from the start. Rapid building was also prompted by the desire to obtain strategic advantage in relation to other competitors, which involved covering as much territory as possible despite limited resources; and by the wish to qualify for land-grants, which were a function of miles built. In these ways some of the differences between American and British railway-engineering can be explained without supposing that the Americans were more interested in saving capital per unit of output. But it is probable that an additional reason was that a higher rate of return on capital was expected of public utilities in the U.S.A. than in Britain, and this would result in lower capital-intensity.

[1] C. Goodrich, *Government Promotion of American Canals and Railroads, 1800–1890* (New York, 1960), p. 153.

[2] The prospects of future expansion also influenced methods of railway construction through its effect on the railways' financial policy. Except in Massachusetts, New York State and Pennsylvania, 'in any majority of cases the (financial) policy pursued is to extract a dividend at the earliest possible moment; to pay that dividend to the last farthing of available surplus over working expenses; and to trust to the increase of traffic for providing, when the emergency may arrive, for the deterioration of the permanent way, and the rebuilding or replacement of worn-out stock. Very few companies, indeed, systematically provide for the renewal of the perishable parts of their property by a reserve fund, regularly put aside out of the annual revenue; and the reason given for the omission is, simply, that by the time permanent way requires renewal, or timber bridges and viaducts replacement, the traffic on the lines will have become so greatly developed that it will be then more advisable to enlarge the capital to the extent required, and replace and renew everything on a more permanent and substantial scale'. (Watkin, *op. cit.* p. 133.)

British canal- and railway-builders tended to project on the basis of 5 or 6 per cent rather than the industrial rate of profit. The Americans required more, and even 1 or 2 per cent more could make a substantial difference in projects in which capital is important in any case.[1]

In a number of instances, therefore, the rapid rate of American investment, in relation to factor-supplies as a whole, was a reason for American interest in methods of economising capital, over and above the reasons previously discussed.

The long-term growth of the market as a result of rising incomes in agriculture and the filling-up of the country is relevant to the argument in a second way: it sustained the incentive to widen capital. According to the argument in an earlier section the level of real wages in American industry was set by the absolute level of investment in relation to labour-supplies, subject to the minimum below which real wages in manufacturing could not go, a minimum determined by earnings in agriculture. The desire to widen capital—because of the protection of tariffs and high transport-costs and the filling up of the country—which determined the level of investment, came up against rising marginal labour-costs, and therefore led to capital-deepening, in order partially to offset the fall in marginal profit-rates. But, within these assumptions, the fall in marginal profit-rates could not be completely offset and would therefore reduce the desire to widen capital which was the impulse which set the whole process in motion. If one introduces technical progress, whether autonomous or induced by the fall in profit-rates, this would help to keep up profit-rates and might raise them, thus maintaining the desire to widen. The relevance of the long-term growth in demand is that it would help to sustain the level of accumulation for normal widening-purposes even in periods when technical progress was sluggish. The growth in demand, that is, contributes to resolving the problem of reconciling the rate of accumulation with the bias towards capital-intensity. This explanation would not be inconsistent with those offered at an earlier stage in the argument but it might render some of them otiose.

[1] In the later 1860's the average rate of interest paid by railway companies in the United States amounted to 8 per cent per annum; in England the rate of interest on long obligations was not over 5 per cent. As a result of this difference it paid to substitute dear (but more durable) steel for iron rails at an earlier stage in England than in America. Abraham S. Hewitt, *Selected Writings*, ed. A. Nevins (New York, 1937), p. 31.

IV

LABOUR-SUPPLIES AND TECHNOLOGY IN THE U.S.A.

In the previous chapters we have explored a number of ways in which factor-endowment is capable of having influenced the composition and rate of American investment. Some of them coincide in general effect, others are mutually exclusive or conflicting; different parts of the argument may illuminate different regions, periods or industries. No attempt is made here to undertake the empirical studies that would be necessary to establish which influences were of most importance in particular times and places. For such a purpose one would need to know the range of techniques facing manufacturers at given points of time and their costs at different combinations of relative factor-prices, the shape of the supply-curves of factors and the demand-curves of products, and the course of industrial profits. On these matters little information is available, and there are two difficulties about using what is available.

(1) The surviving data are the *results* of what happened. What we need is not a series of labour-costs per unit of output but the expected cost, at a number of points of time, of acquiring additional labour; perhaps the nearest approach would be piece rates over a period of reasonably unchanging techniques.

(2) The relative costs of techniques were influenced by forms of industrial organisation as well as by relative factor-prices. It is difficult, for example, to compare the relative costs of old and new methods in the cotton textile industries of the two countries, where most data are available, because, while the old methods were carried on in the same way in both countries, a much higher proportion of the new equipment was installed in *combined* weaving- and spinning-concerns in the U.S.A. than in the U.K.

Moreover, one cannot by direct inspection of technology establish the part played by factor-scarcities, compared with other influences, in determining the technology of the two countries. Even if one were to find a much higher proportion of labour-savers among American techniques, as casual inspection suggests one would, this would not necessarily show that labour-scarcity was a more powerful stimulus to invention than scarcity of other factors. For the explanation might be that Americans were intrinsically better inventors than the British and, since they were influenced in their inventions by American factor-endowment, a high proportion of new inventions consisted of those which suited American circumstances better than the English. This is to say, it might be the case that a shortage of natural resources *per se* was just as great a stimulus to invention as was labour-shortage but that the American inventors were better at responding to their stimulus than the British to theirs. Moreover, the effects of an invention are not always closely related to the purposes for which it was initially designed, and a number of inventions are not at all inspired by factor-scarcities and are so productive that they are adopted irrespective of factor-proportions; the economic character of technology might therefore reflect heavily the different extents to which such inventions were made use of, and they might neutralise—or for that matter reinforce—the effect of factor-scarcities.

All that is attempted therefore in the immediately following pages is to speculate about the *a priori* arguments in the light of a small number of facts.

In the eighteenth century much of such industry as existed in the U.S.A. was conducted by small scattered manufacturers, drawing their labour from, and selling their goods in a very limited area. They were not unlike the small businesses of medieval Europe—the early American towns had a number of medieval practices such as regulations against forestalling and regrating—and, confined to a small circle of customers by lack of communications, some of them extended horizontally, giving more services to a given group of customers, and some went in for part-time farming and land-speculation. Until the local labour- and/or market-possibilities were exhausted, they enjoyed high profits; but, beyond this point, they suffered a sharp fall in the marginal rate of profit on capital. Where the goods were bulky—ploughs, furniture, doors, etc.—there might be virtually no possibility of expanding the market, once local requirements were met, because of the high cost of transport; and inelasticity of the demand for products might precipitate a severe fall in marginal profit-rates before the imperfection of the labour-market had begun to exert serious pressure. Where the goods were more valuable in relation to their bulk, demand for them was more elastic and pressure on the rate of profit was more likely to come from exhaustion of the local labour-supply.

There were thus considerable differences in the rate of profit on capital—average and marginal—in different localities, according to the local elasticities of labour-supply and of product demand and the stage of development. The need to reduce costs per unit of output in order to mitigate the fall in the marginal rate would set in at different times and would assume different forms according to whether the fall was brought about by rising marginal cost of labour or by falling product-prices.

Some local manufacturers in districts protected by distance had probably not reached the point of sharply falling returns even by the end of the eighteenth century. Other manufacturers reached the point early, continued to expand output, exploited fully the known means of economising labour and occasionally invented new methods of doing so. There was an impressive amount of labour-saving invention in America before the end of the eighteenth century; for example, Oliver Evans's milling-machinery, which introduced a primitive form of production-line and halved the labour required, the Perkins machine for the manufacture of pins (invented in 1790 and patented in 1795), and numerous nail-making machines.[1] Other manufactures, again, reached the end of easy expansion early and were content not to press beyond the possibilities of

[1] G. and D. Batho, *Jacob Perkins* (Philadelphia, 1943); *Oliver Evans* (Philadelphia, 1935), pp. 11ff. In 1812 Thomas Attwood, before the Committee on the Orders in Council, said that in the last three or four years the U.S. had succeeded in the manufacture of nails by 'a kind of stamp machine, not altogether a perfect machine, but a machine capable of being made perfect, and capable of superseding as I have been informed, all mechanical labour . . . there shall be no more nails made with hammers'. *Evidence of Petitions before the House of Commons on the Orders in Council*, P.P. 1812, III, p. 9. 'In America . . . the dearness of labour, coupled with the large demand for nails to be used in the erection of timber building' stimulated attempts to devise machinery for replacing hand-labour—no less than twenty-three patents for nail-making machines had been granted in America before 1800. Richard B. Prosser, *Birmingham Inventors and Inventions* (Birmingham, 1881), p. 70. Gallatin (Report on Manufactures, pp. 436–8) lists a number of recently-invented mechanical devices, including machinery for manufacturing wool and cotton cards (Amos Whittemore, 1797), Stowel's patent for a machine to produce wood screws (1809), Johnathan Ellis' machine for cutting nails. 'In old countries, nails are forged; here they are cut.' It is not clear whether Whittemore was American or English: see p. 121 f.n. 2 below.

the local market; this does not mean that they ceased to expand but only that they did not expand faster than the increase in agricultural incomes and/or labour-supplies in their locality. Which of these types of manufacturer was the most characteristic would require very detailed investigation to establish. But it is probable, on general grounds, that many American manufacturers around 1800 were content not to press beyond their local possibilities. In this period the available techniques were primitive and the scope for movement into more capital-intensive techniques was limited; a minimum of technical knowledge is necessary before the search for labour-saving devices can be widely fruitful. Where the difficulties of expansion beyond a certain point were very great, a manufacturer might very well decide that expansion was not worth the trouble; especially since the small manufacturer of those days had an alternative use for capital in the purchase of a farm.

Moreover the penalties for non-expansion, or for expansion with equipment which was inappropriate to the country's factor-endowment, were not severe. So long as labour- and product-markets were highly imperfect, though there was a great need to mechanise if expansion beyond some point were not to lead to a sharp fall in the marginal rate of profit, the penalty for not expanding was simply that total profits were smaller than they would otherwise have been.

Thus many American manufacturers at the close of the eighteenth century were poised at the point at which average profits were high but where marginal profit-rates threatened to fall rapidly. They were using labour as economically as possible with existing techniques and were on the look-out for new methods of doing so, but only in a limited number of cases had they been so successful as effectively to defer the point of sharply falling returns.

This was a situation which was capable of producing new methods, even without the operation of any forces external to the American economy. A sharp increase in demand, for example, might provide just the additional force necessary to break through the technical difficulties which impeded the replacement of men by machines. Thus the government contracts for arms to Whitney (1798) and North (1799) provided them with the impulse to undertake revolutionary changes in their manufacture; there was nothing to compare with these changes in the contemporary British manufacture of small arms, despite the fact that it too was meeting a great increased demand.[1]

Moreover, the penalties for not adopting capital-intensive techniques increased as communications improved, and as labour- and product-markets became less imperfect. If *some* manufacturers expanded output by adopting capital-intensive methods and in so doing raised the labour-costs of their competitors, those who did not expand, and still more those who expanded with inappropriate equipment, became high cost producers and lost their share of the market. This was true even if the adoption of capital-intensive techniques did not place manufacturers in a better position to take advantage of the possibilities of technical progress. But if they were so placed, they might threaten the product-markets even of competitors who still enjoyed a monopsonist position in their local labour-markets, for these, lacking the inducement to adopt capital-intensive techniques, ultimately fell behind in technology. Since imperfections were relaxed more rapidly in the product than in the labour-markets the last was probably the situation of most practical importance.

It is quite possible that most of the mechanisation which impressed English observers in the 1850's was due to autonomous developments of this sort. The system of interchangeable parts was certainly such an autonomous invention, as were many of its later adaptations, for example the invention of the sewing-machine. But in the early nineteenth century there were also important external influences. The first was the new technical knowledge produced by the classic English industrial revolution. As we shall argue later, the inventions in the English textile industry in the eighteenth century were principally the result of attempts to save labour. Some of them saved labour at the expense of some increase in capital-costs, and even when the fall in total inputs per unit of output was so spectacular that the new method was the most advantageous for any feasible range of factor-prices, the fall in labour-costs was greater than in capital-costs and the advantage was most evident and decisive where labour was dear. The new methods in textiles were therefore well suited to the American circumstances and they

[1] C. McL. Greene, *Eli Whitney and the Birth of American Technology* (Boston, 1956), pp. 101–2. S. M. D. and R. North, *Simeon North* (Concord, N.H. 1913).

were rapidly adopted. Where, as in the English primary iron industry, the main motive of the new methods had been fuel economy, the course of adoption was different. Whereas in England charcoal had been replaced by coke first in the furnace and then in the refinery, and the improvements in rolling were essentially ancillary, in the U.S.A. the order was reversed; rolling, which saved labour, was adopted first, and the use of coke in the blast furnace last. The different speed of adoption of the new technology, in different parts of the world, is commonly explained by reference to differences in the quality of entrepreneurial ability. But at least as important in explaining the speed of its reception in America is the character of American manufacturing when the new technology became available.

It is true that in the early phase of development the Americans, in adopting the new techniques, were forced to rely on American-made machines and that these were dearer than machines made in England. At the very start, indeed, the disparity may have been greater in the cost of machines than in the cost of labour, and if this was the case the high level of American wages would not in itself explain why Americans should adopt more capital-intensive techniques than the English. But in America, during a boom, the supply of machine-makers (skilled labour) was more elastic than the supply of unskilled labour because of the heavy demands on the latter to open up the new areas; that is, the supply of machines was more elastic than the supply of labour. In England on the other hand, during a boom, because of the reserve army of agricultural and ex-agricultural labour, the elasticity of supply of skilled workmen was less than that of unskilled, that is the elasticity of supply of English labour was greater than that of English machines. In these circumstances, since most innovations were made in the boom, there would have been a greater bias towards capital-intensity in America than in England, even if the average cost (average over boom and slump) of American machinery compared to American labour, was higher.

The removal of the prohibition on English exports of machinery introduced a second new element. It became easier for American manufacturers to buy machines, including machine tools, made with the cheaper English labour. The prohibition of the export of machinery was not removed until 1843, but the prohibition had been in practice relaxed by licence or evaded by smuggling for some time before, and though the chronology of the relaxation requires further study there was by the end of the 1820's a substantial export of machinery from England. By 1841 licences were given for nearly everything except spinning and weaving machinery. Though the price of English machines in America was higher than in England, the disparity was not so great as between the cost of labour in America and in Britain. The Americans now had an incentive to use more machines than they had previously—to replace dear American labour by machines made with cheaper English labour; they had an inducement to install the English machines existing at the time the prohibition was relaxed, and any new ones which the English invented thereafter. The removal of the prohibition, that is, produced a once and for all increase in the bias towards capital-intensity. Moreover, the possibility of importing machines from England, where the labour-supply was more elastic, made it easier to mitigate the effect on profits of a given rise in marginal labour-costs. A given rise in the marginal cost of American labour was now a *greater* inducement to shift to capital-intensive methods because it was possible to import machines from England where the cost of labour and therefore of machines had not risen. America developed her own cotton-textile machinery, and the machines for producing small arms, clocks and watches, and sewing-machines, but there was some import of heavy machine tools from England, and of machines for the manufacture of metal toys; and most of the machines in the worsted industry were of European origin.[1]

At the same time there were probably also changes in the elasticity of labour to the individual American manufacturer. As transport improved and the country was settled, the labour-market in the eastern states became less imperfect; more manufacturers were able to draw on a common reserve of labour and fewer were in the position where, beyond a certain point, they could virtually not obtain any additional labour. As against this the rate of capital accumulation was probably faster than it had been in the eighteenth century, since the

[1] F. W. Taussig, *Some Aspects of the Tariff Question* (Cambridge, Mass. 1915), p. 343. English machinery for the manufacture of brass buttons was imported into the U.S.A. Lathrop, *The Brass Industry* (Shelton, Conn. 1909), p. 89. The main English exports of machinery to the U.S.A. before 1841 were the superior lathes and planing machines. P.P. 1841, VII, p. 21, Qu. 217.

introduction of the new techniques mitigated the tendency of the rate of profit to fall and probably led to some increase in the rate, and since the internal market for American manufactures was also expanding more rapidly than it had done earlier. It was easier to get a given amount of labour, but more labour was required.

The Embargo Act of 1807, the Non-Intercourse Acts and the War of 1812 almost certainly did more than compensate for the net effect on profits of the higher costs of American labour. The *ad valorem* duties of between 20 and 25 per cent imposed on most manufactured goods by the tariff of 1816 were probably not sufficient to do this if contemporary estimates of the disparity between English and American labour-costs were correct.[1] But the Tariff Act of 1824 raised duties to a level which for certain commodities afforded adequate compensation for dear labour at the levels of output then prevailing. Because of imperfections in the product-market within America, American manufacturers were enabled for some time to enjoy a higher rate of profit on capital than the English; there must have been places in America where the market was supplied by imports and by a local monopolist who could charge up to the c.i.f. cost of imports plus the tariff.

But the tariff, though initially it gave more than adequate protection for the high cost of American industrial labour, was not adjusted in the years that followed to take account of any rise in marginal costs of American labour. The Compromise Act of 1833 gradually undermined the tariff of 1824 and by 1842 the rates of duty were lower than in 1824. There was also a long-term fall in freight charges. These changes probably more than offset the increase of the incidence of the tariff where the rates were specific. The competition of imports set a limit to the extent to which American manufacturers could alleviate the effect of rising labour-costs on the rate of profit on their capital by raising product-prices. On this reading of the situation, the free-trade interests of the southern states, by restraining tariff increases, may have accelerated rather than hindered the mechanisation of American industry.

The pressure to invent new methods obviously varied in different parts of the economy. It was greatest where the exist-

[1] J. Potter, 'Atlantic Economy, 1815–1860', in *Studies in the Industrial Revolution*, ed. L. S. Presnall (1960), pp. 253–5.

ing range of techniques afforded least possibility of escape from dear and inelastic labour-supplies; there were some activities in which organisational improvements by themselves, without any changes in technique, could afford the manufacturer considerable relief; in some cases some time elapsed before the labour-saving possibilities of the technology taken over from England were fully exploited, and before the need to improve upon it became pressing; in other sectors, technical innovation was a condition of expansion from the start. The rate of innovation actually achieved in response to a given incentive depended on the technical possibilities—the nature of the product, the processes and raw materials involved—and on the nature of the market: broadly speaking, labour-saving machinery could most easily be devised for products made in large quantities to a single pattern.

Thus the point of time and the extent to which the need to save labour bore fruit in technical progress varied from activity to activity. But it may reasonably be conjectured that by the 1820's and '30's pressure was building up, within a number of fields, to invent new techniques. A witness before the Parliamentary Committee on the laws relating to the export of machinery said that 'The greatest portion of new inventions lately introduced to this country have come from abroad; . . . by that I mean not improvements in machines, but rather entirely new inventions. There are certainly more improvements carried out in this country; but . . . the chief part, or a majority, at all events, of the really new inventions, that is, of new ideas altogether, in the carrying out of a certain process by new machinery, or in a new mode, have originated abroad especially in America'.[1] In 1833 an English business man, appearing before the Select Committee on Manufactures declared that 'of

[1] 1st Report of S.C. on the Exportation of Machinery (1841, VII, p. 117). This witness, a Manchester machine-maker, mentioned the following new labour-saving processes introduced from the U.S.A.: a machine for making nails and tacks (English patent, 1810), a card-making machine (introduced about 1797 by its inventor, Whittemore of New York, and patented in England in 1810), Perkins' engraving process, a reed-making machine, a shearing machine [*ibid.* pp. 118, 129]. A number of witnesses before this committee testified to the importance of labour-scarcity as a stimulus to American invention. Thus an American witness who had been engaged in the erection of machines in America asserted, in answer to a question whether American invention had been turned chiefly to the construction of labour-saving machines, that 'American invention has been applied almost exclusively to the production of such machinery. In consequence of wages being very high in that country, the object of manufacturers has always

all the improvements of great importance (in cotton textiles) that have been made for the last two years the invention has been in America. ... The improvements in the machinery itself (as distinct from improvements in its working) ... have been the inventions of America'.[1] There is no reason to doubt contemporary attribution of this inventiveness to labour-scarcity. One of the Commissioners sent to the U.S.A. by the English Board of Ordnance in 1854 commented that, 'On account of the high price of labour, the whole energy of the people is devoted to improving and inventing labour-saving machinery'.[2] The Jurors who reported on the New York Industrial Exhibition wrote to similar effect: 'The very difficulty in procuring human labour, more especially when properly skilled and disciplined ... appears to have stimulated the invention of the few workers whose energies and skill were engaged in the early development of manufactures; and to this very want of human skill, and the absolute necessity for supplying it, may be attributed the extraordinary ingenuity displayed in many of those labour-saving machines, whose automatic action so completely supplies the place of the more abundant hand labour of older manufacturing countries'.[3]

These statements may well exaggerate the number of important new inventions in the first half of the nineteenth century which were of American origin. The absolute amount of mechanical experience and ingenuity were greater in England, so that it would be surprising if most seminal inventions had *not* been of English origin. The point is that the American inferiority on this score was offset, to an increasing extent, by the specific incentive to invent which their factor-endowment provided.

In agriculture, because of the abundance of land and the scarcity of labour, the Americans had from early times concentrated on extensive cultivation. 'The English farmer' wrote Washington to Arthur Young 'ought to have a horrid idea of

been to introduce and employ labour-saving machines ...' (Qu. 3024). 'In America ... labour has hitherto been dear, comparatively speaking, from a want of labourers; parties, not having access to the machine makers of this country (that is England) for supply of their wants, have themselves set about to make a machine in the readiest mode to accomplish that which they required; they have been untrammelled by predilections in favour of a machine already in existence' (Qu. 1545).

[1] P.P. 1833, VI, Qu. 640, 647.

[2] P.P. 1854–5, L, p. 51.

[3] P.P. 1854, XXXVI, p. 12.

the state of our agriculture or of the nature of our soil, when he is informed that one acre with us only produced eight or ten bushels. But it must be kept in mind, that where land is cheap and labour dear men are fonder of cultivating much than cultivating well'.[1] But since there was so much land to cultivate, Americans also had a very strong incentive to develop machines which would enable farmers to cultivate a larger area. The alternative was to leave land uncultivated. It was said of Illinois in 1857 'all grain is here cut by machine. Cradles are out of the question . . . If grain is too badly lodged to be so gathered it is quietly left alone. . . . This work done by machinery is not very much cheaper than it could be done by hand, but the great question is—where are the hands to come from?'.[2] In England, where land was scarce but agricultural labour abundant, farmers were principally concerned with raising output per acre rather than output per man. It is therefore not surprising that it was in agricultural machinery that America first established an international reputation for her machines, and in this field she ultimately made the most striking advances. A British patent for a reaper was taken out in 1799 and the first American patent in 1803; but the several technical difficulties, particularly the difficulty of making an efficient cutting-bar were overcome first in America.

This does not imply that there were no land-saving improvements in the U.S.A. and no labour-saving improvements in England. Improvements in American agriculture took the form partly of regional specialisation achieved by trial and error. Farmers found from experience what types of crop did best and where, and land was sorted out into its most appropriate uses according to its natural qualities, market possibilities, and existing ratio of available labour to land (which while low everywhere was not uniform). The general effect of this process was probably some increase in output per acre. Other American improvements—steel ploughs, windmills, new strains of crop—were important because they enabled new land to be brought into cultivation, and they were therefore in a sense, land-saving improvements comparable with drainage in

[1] E. G. Wakefield, *op. cit.* pp. 11, 51.

[2] *Iowa Farmer* (1857) quoted by C. H. Danhof, 'Farm-Making Costs and the "Safety-Valve": 1850–1860', *Journal of Political Economy*, XLIX, No. 3 (June 1941), pp. 348–9.

England. In England on the other hand, the agricultural improvements associated with enclosures must have caused some increase in productivity per head, for example by facilitating a more even spread of work over the year; and very considerable improvements were made to English agricultural machinery.[1] But of the broad general contrast between the two agricultural technologies, there can be no reasonable doubt.

That shortage of labour and abundance of land had always provided an incentive to mechanise American agriculture is obvious. But it is quite conceivable that labour-scarcity exerted its stimulus as powerfully in industry. The fact that, in agriculture, wages, profits and rent so often accrued to the same person meant that, as capacity expanded, the change in the cost of labour was less visible than in industry, and that additional labour was often forthcoming simply from the greater efforts of the farmer and his family. For these reasons it is possible that, for any given rate of accumulation, wages in terms of product rose more in industry than in agriculture, or, at least, were more evidently seen to rise, thus providing a more marked incentive to adopt the more capital-intensive methods.

In cotton textiles, the possibilities of improving upon the methods borrowed from England and of adapting them to American circumstances proved considerable, principally because cotton, a tough fibre, could stand up to mechanical operations, and because it was possible to standardise cotton fabrics. Of the years 1814–24 it has been said that 'of all the processes involved in cotton manufacture, only carding and drawing failed to be radically improved in this decade',[2] and the rate of technical innovation continued to be high down to the mid-'40's. The most important adaptations and inventions —the ring-spindle, the Goulding automatic roving machine, the self-acting temple, and the various self-stopping devices in case of breakage—were in large measure due to the dearness and instability of the available labour.[3] In weaving, mechanisa-

[1] J. A. Ransome, *The Implements of Agriculture* (1843).

[2] Gibb, *op. cit.* p. 36.

[3] W. F. Hayes, *American Textile Machinery* (1879), pp. 31–32. The adoption in the U.S.A. of the ring-spindle in preference to the mule was due, among other things, to the fact that it was more productive per operative and used less skilled labour. 'In Great Britain a class of skilled cotton mill operative, almost an hereditary class, provides an adequate supply of skilled mule-spinners. In America on the contrary, the cotton manufacturers have always had to adapt themselves to a class of shifting and unskilled workmen' (Copeland, *op. cit.* pp. 72–3).

tion proceeded rapidly after F. C. Lowell designed his power-loom in 1813; and the early adoption of the power-loom enabled the Americans to make a continuous series of improvements, more far-reaching than anything in England: for example William and George Crompton's loom in 1831, adapted to cassimeres in 1837–8 and culminating in an automatic loom in 1842.[1]

The woollen and worsted industries afforded fewer opportunities of mechanisation than cotton simply because wool is less amenable than cotton to machinery of the quasi-automatic kind: '... the nature of wool and the yarn spun from it, as well as the more diversified character of the fabrics, stand in the way of any sweeping application of the methods (ring-spinning and automatic looms) ... which have proved of such far-reaching effect in the weaving of cottons'.[2] And in technical advance in woollens, the English had certain advantages. The English retained the less capital-intensive mule-spindles; because of the quality of the fibre, mule-spindles were better suited to wool than were ring-spindles, and the English were therefore in a better position to make such technical progress as was made in the spinning of wool. In this section of the industry, according to Professor Cole, the adoption and development of power-driven machinery before 1830 was not appreciably more rapid than in England, and between 1830 and 1870 America failed to keep pace with England, particularly in the use of self-actor mules.[3] But in the weaving section of the industry, the Americans adapted the power-loom to woollens much more rapidly than the English, and they also made important advances in carding and in the finishing processes. Taken as a whole, this American industry, which started about 1800 with techniques entirely taken over from England, made more rapid technical progress in the following thirty years than the corresponding English industry, and also

[1] Taussig, *Some Aspects of the Tariff Question*, *op. cit.* p. 293. In carding and spinning, especially in the former, Montgomery thought that the U.S.A. was much behind Britain, but in weaving by power 'the Americans have in every respect equalled, and in some things surpassed, anything I have yet seen either in Glasgow or Manchester' (p. 82).

[2] Taussig, *op. cit.* p. 357.

[3] A. H. Cole, *The American Woollen Manufacture* (Cambridge, Mass. 1926), I, pp. 120, 360.

equipped itself mainly with machines made in America.[1] The American worsted industry, by contrast, was introduced into America much later, after 1845, and 'a complete equipment of well-developed, quasi-automatic machinery was originally borrowed from abroad, and, since its introduction here, it has undergone scarcely any important improvement in type'. Most of the machines in the American worsted industry were of European origin. Views differ about the reasons for the contrast with the woollen industry,[2] but is it not possible that labour was less scarce when the worsted industry was introduced, and that in these circumstances the imported techniques were sufficiently labour-saving?

But in the iron industry there was no way of offsetting the high wages until the development of bituminous coal supplies in the 1870's and the development of the Lake Superior iron region. Cort's processes of puddling and rolling which for three-quarters of the nineteenth century were the methods of manufacturing wrought iron, depended largely on labour, and technical progress did not provide a markedly labour-saving method until the Bessemer process. Here the abundance of fuel—as well possibly as consumer resistance—reduced the need for the Americans to adopt the techniques which proved to have the greatest technical possibilities.[3]

But the fields in which the effect of labour conditions on technique had the most important long-term consequences were those to which it was possible to apply the method of interchangeable parts, the 'American System' as it became known. Its original development obviously owed a great deal to the intelligence and ingenuity of Whitney and North and to the particular circumstances of the industry in which they worked—since the demand for arms was price-inelastic and the costs of innovation could, to some extent, be passed on to the government. It is the sort of idea which might conceivably have struck anyone sufficiently bright, whatever the conditions of labour-supply, as it struck Brunel and Samuel Bentham in England about the same time. But it was in fact devised to overcome a shortage of skilled labour and its practical development and application to products other than small arms was undoubtedly the result of labour-scarcity.[4]

[1] *Ibid.* pp. 86, 133.

[2] *Ibid.* p. x, 362.

[3] Taussig, *op. cit.* p. 118.

[4] Strassmann, *op. cit.* pp. 146–7.

The method proved applicable to the manufacture of a wide range of products: wood-screws, woodworking, nuts, bolts, locks, clocks, watches (in 1848), agricultural machinery, footwear (mechanised after 1850 with separate parts for each machine), and to new products like the typewriter (in the 1840's and '50's) and the sewing-machine (the first completed machine for general use was patented by Elias Howe, Jnr of Cambridge, Mass. in 1846), as well as to locomotives (after 1860) and a whole series of motors. Perhaps its most technically interesting application was in the invention, by Hollerith, of a machine for tabulating statistical returns.[1] The method was also to some extent applicable to the manufacture of machine tools, and therefore reduced the price of machines. But even where this was not the case, the manufacture of final products by the method of interchangeable parts gave Americans an advantage in the production of machine tools.

A large part of American industrial progress in the nineteenth century was due to the rapidity of technical advance in machine tools. Most of the general machine tools had been invented by British engineers between 1775 and 1850—boring machines, engine lathes, planers, shapers, steam hammers, and standard taps and dies; and in trustworthy general-purpose tools the English held their own up to and beyond the end of the nineteenth century.[2] But the most important new machine tools, particularly the milling machines and turret lathes, were developed in America and were adapted to very specialised uses. They were more specialised because the parts they were used to make were more specialised; and because they were more specialised they became cheaper. Whereas the English general-purpose tool had to be strong enough to survive the most exacting of the many uses to which it might be put—and was therefore unnecessarily strong for many of its uses—the more specialised American tool could be made just strong enough for its restricted use. And the specialisation of the machine tool and the breakdown into parts of the final product interacted.

American excellence in the production of machine tools did not become evident until after the 1850's and later. In 1839

[1] C. H. Fitch, Report on the Manufactures of Interchangeable Parts, in *U.S. Census, 1880*, volume on manufactures; Levasseur, *op. cit.* pp. 59–60.

[2] Chapman, *Foreign Competition*, *op. cit.* pp. 125–6.

the Locks and Canals Company of Lowell, the largest of the non-specialised manufacturers of textile machinery in America, imported from Whitworths a planer and a 'self-acting upright drilling and boring machine'.[1] The report on machinery in the 1850's thought that, in point of quality, American tools were inferior to the English: 'as regards the class of machinery usually employed by engineers and machine-makers, they are on the whole behind those of England'. 'But', the report continued, 'in the adaptation of special apparatus to a single operation in almost all branches of industry, the Americans display an amount of ingenuity, combined with undaunted energy, which as a nation we should do well to imitate'.[2] Even in the 1850's the Americans had, for a number of operations, machine tools more specialised than those available in England particularly in woodworking and small arms. And the rapid emergence in the 1850's and '60's of firms of specialised machine-tool builders multiplied their number, until by the 1880's, it has been said, the price of American machine tools had fallen to half that of the equivalent British tools.[3]

So far as the machines themselves are concerned, the best English opinion in the 1840's seems to have believed that where the same type of machine existed in both countries the English version was better and cheaper. James Montgomery thought this of textile machinery (with the possible exception of looms) and metal-working machines. 'The number of machines taken from America', wrote Montgomery, 'for which patents have been obtained in Great Britain, has led many to suppose that the Americans have attained to considerable perfection in labour-saving machinery. This, however, is not the case; nor do I think that, in this respect, they are at all equal to the British'.[4]

[1] Gibb, *op. cit.* p. 81.

[2] P.P. 1854–5, L, p. 578. The Committee emphasised 'the adaptation of special tools to minute purposes, in order to obtain the article at the smallest possible cost' (p. 630).

[3] Strassmann, *op. cit.* p. 117. 'The application of machinery to the working of wood enables the manufacturers to get up the ordinary ploughs at inconceivably low prices' (P.P. 1854, XXXVI, p. 445).

[4] Montgomery, *op. cit.* p. 109. One witness before the S.C. on the Export of Machinery (1841), who from the tenor of evidence was clearly prejudiced against American achievements, estimated that American machinery for cotton spinning in 1832–3 cost nearly three times as much as the English (P.P. 1841, VII, p. 129, Qu. 1816), partly because American iron and steel were dearer and partly because American machine manufacturers were 'not systematic in their work. Here . . . we

Even at the mid-century, however, English comment—which was more technical than economic—may not sufficiently have taken into account the different requirements of the markets in the two countries. Montgomery's complaint was that American textile machines could be less easily adjusted to meet changes in market requirements, but this may simply reflect the greater homogeneity of the American market.[1] He also criticised the flimsy, less durable nature of American machinery. But the more exacting standards of construction in England may well have been uneconomic and characterised by 'needless weight, unnecessary finish and complicated movements'; and the fact that American machines were less durable may have meant that the period of physical obsolescence approximated more closely than in England to what proved to be the period of economic obsolescence.[2] And from the 1860's on, as a result of the use of interchangeable parts in the manufacture of machines and the use of cheaper machine tools in the manufacture of these parts, the price ratio of superior to inferior machines was narrowed in the U.S.A. more rapidly than elsewhere.

We should expect to find that factor-endowment influenced not only the composition of investment in comparable in-

take separate branches, and we can turn more out; there a man is a turner, a filer, and a fitter-up'. A more balanced (American) witness estimated that English cotton and wool machinery was superior (Qu. 2934, 2995) and that in the finer descriptions of spinning machinery the American machines were 60 per cent dearer than the English, in coarser machines 40–50 per cent (Qu. 3002). On the other hand, William Vickers, the Sheffield steel manufacturer, thought that the Americans had so far improved their machinery that in many points they excelled the English, and that even if the prohibition on the export of English machines were removed, the English would capture very little of the American market in machines 'except in some articles which contain a large quantity of labour' (Qu. 4287, 4288). For a comparison of English and American agricultural machinery, see P.P. 1854, XXXVI, pp. 429ff.

[1] Montgomery, *op. cit.* pp. 109–10. In Britain 'every (cotton textile) machine is so constructed that all its parts can be adjusted with the greatest accuracy, to suit the various qualities of cotton, or whatever kinds of goods may be wanted: and consequently, the manufacturers can easily arrange and alter their machinery, so as to make just such goods as will for the time being suit the market. But here (U.S.A.) they can make only one kind of goods, whether they suit the present demand or not, as they have not the same facilities for changing the style of their goods, so as to take advantage of variations in the market'.

[2] D. L. Burn, *op. cit.* p. 299. For a typical comment see the Committee on Machinery: 'All the machines used for woodwork in the United States are roughly constructed and would not bear comparison in stability and appearance with the highly-finished iron machinery of England'. (P.P. 1854–5, L, p. 75.)

dustries in the two countries but the composition of their industrial sectors. The Americans had a comparative advantage in branches of industry with a high content of those raw materials which were cheap in relation to other factors and had a low transport content, for example, in the making of wooden ships. In addition, the reasons for supposing that labour-scarcity gave a capital-intensive bias to American investment within a given industry would also lead one to expect that Americans would tend to specialise in those industries, or branches of industry, where the possibilities of compensating for the high cost of American industrial labour by the adoption of capital-intensive techniques were most promising. To the extent that technical progress and the propensity to save out of profits were functions of capital-intensity these would, other things being equal, be the most rapidly expanding American industries. By contrast one would expect the English to specialise in labour-intensive industries.

How far the industrial structure of the two countries in fact reflected their factor-endowment could only be settled, even in general terms, after a detailed investigation which it is not the intention of this essay to undertake. But it is possible to identify very broadly by reference to the tariff, the fields in which the American search to compensate for high labour-costs was most successful. It is an extremely rough measure, for the tariff also reflects the political power of various interests, and some industries needed protection simply to compensate for the high price of inputs brought from other protected industries. But the degree of protection needed by an American industry is probably the best measure we have of the adequacy of the labour-saving methods available in any given period to offset the higher American wages.

But though the tariff gives some clue to the fields in which the attempts to compensate for dear American labour had been most successful, it is extremely difficult to say what success these attempts had achieved as a whole by the mid-century. Nathan Appleton, writing in 1831 about labour in the cotton-textile industry, argued that labour-costs per unit of output were lower in the U.S.A. than abroad.[1] And Carey, in a more general treatment of wage differences but one which rested

[1] Nathan Appleton, *Introduction of the Power Loom* (Lowell, 1858), p. 30. For a similar view see P.P. 1833, VI, Qu. 2616, 5385.

on the statistics of the cotton-textile industries, concluded that 'any difference in (money) wages that may exist between England and the United States, must arise out of its better application in the latter'.[1] But if this were so it is hard to see why American manufacturers needed the tariff. And even in an industry like the manufacture of small arms, which was indisputably much more high mechanised in the U.S.A. than in England, it is not clear that in the 1830's and '40's these more mechanised methods did a great deal more than compensate for dear American labour. One of the few products for which information about productivity exists over a long period of time is gun barrels. The average number of barrels welded per man per year rose more or less continuously from 1806, but the rise was not very marked until the 1840's and only became spectacular in the 1860's. The increased productivity of labour was accompanied by a fall in labour-costs per barrel, but the fact that there was no reduction in the cost of the finished weapon suggests strongly that there was an offsetting increase in capital-costs per unit of output; and this is what the historian of the industry implies. 'Mechanisation, however great its value in improving the quality of arms failed in this case (the Springfield Armoury) at least to bring a marked reduction in costs, and perhaps even increased them through added requirements for insurance and depreciation charges'. Mechanisation improved the quality of small arms—and in this field small margins of superiority are obviously of critical importance since they make a difference between winning or losing engagements—and also enabled more guns to be produced with a given labour-force. But the increase in capital-costs necessary to achieve this seems, at least until the 1840's, to have offset the fall in labour-costs.[2]

Moreover, in some industries—textiles, mining and possibly light engineering—labour-scarcity compelled entrepreneurs to adopt methods which were capital-intensive but the principal effect of which was not to increase *per capita* output but to improve working conditions. Higher productivity might

[1] Carey, *op. cit.* p. 88. See also Gallatin (*op. cit.* p. 430): The high price of labour 'is now, in many important branches (of industry) nearly superseded by the introduction of machinery'; and Watkin (*op. cit.* p. 125): 'Although daily wages are higher in the States than with us, I imagine that the work of (railway) construction is done as cheaply'.

[2] Deyrup, *op. cit.* pp. 132, 246, Appendix D, fig. 2.

result as an incidental by-product, but the main purpose and effect of such improvements was to lighten the load and lesson the toil of human labour. Immigrants' letters sometimes refer not so much to the effect of machinery in maintaining higher wages as to its use to ease the burden of the worker.

One would suppose that inventions such as the typewriter and the sewing-machine were so labour-saving that they did very much more than compensate for dear American labour. But there is a case for suspecting that until the 1840's mechanisation and invention did not often do a great deal more than compensate for the dearness of labour. For the national income estimates—though they are, of course, extremely fragile and do not in any case purport to measure improvements in the burden of work—seem to show that the course of income per head in the first half of the nineteenth century was much the same in England as in America.[1]

Something ought also to follow from our analysis about the distribution of income within the industrial sectors of the two countries. Any increase in industrial capacity redistributed American national income in favour of wage-earners. The shift to the more capital-intensive techniques moderated the redistribution, but so long as it did less than compensate for the higher cost of American labour it would not restore the previous distribution. For a given degree of technical progress the scarcity of labour in the U.S.A. would ensure a rise in the share of labour greater in America than in Britain. If American manufacturers had a higher propensity than the English to save, this would further reduce their share, through the effects on higher wages which could not be entirely offset by greater capital-intensity. In the early ninteenth century therefore we should expect the share of labour within the industrial sector to have been higher in the U.S.A. than in England.

[1] R. F. Martin, *National Income in the United States* (National Industrial Conference Board, New York, 1939), pp. 6, 7: 'very little actual economic advance *per capita* appeared in the first half of the nineteenth centruy'. E. M. Ojala, *Agriculture and Economic Progress* (Oxford, 1952), p. 26. M. Frankel, working back from the known data about labour productivity and making plausible assumptions about relative rates of increase, suggests that British and American labour productivity in manufacturing industry were about equal in the 1830's, and that America began to move ahead sometime before 1870. *British and American Manufacturing Productivity* (University of Illinois, 1957), pp. 26–29. For criticism of Martin's estimates see S. Kuznets, *Income and Wealth of the United States* (Cambridge, 1952) pp. 221–239.

If, however, the incentive to mechanisation came principally not from the general scarcity and inelasticity of labour, but from imperfections in the labour-market, it is possible to imagine a situation in which a bias towards capital-intensity of investment would be compatible with a higher share of profits in the American industrial sector than in the British. But though it is possible to conceive such a situation a scarcity of labour explanation would in general lead one to expect that the share of labour would be higher in industrial America than industrial England.[1]

This discussion is not intended as a complete account of the influences which were favourable to American industrial development in the first half of the century. There were other influences of a less strictly economic kind than the factor-endowment of the country.[2] One possibility is that American

[1] There are no figures for the distribution of income in the nineteenth century in the industrial sectors. The national figures, quite apart from their fragility for this period, cover too great regional and occupational diversity to be of much use. Such as they are, they do suggest that the share of labour in the national income around the middle of the century was significantly higher in America, though this may well be due to the relative importance of agriculture.

Percentage of Labour in National Income

U.S.A.		U.K.	
1850	80	1843	63
1910	74	1913	62·5
1938	66·5	1938	62·5

J. Tinbergen and J. J. Pollack, *The Dynamics of Business Cycles* (1950), p. 37.

It has been argued that the constancy of relative shares in the United Kingdom between 1880 and 1914 suggests that capital was not easily substitutable for labour (J. R. Hicks, *Theory of Wages* [1932], p. 123). By the same token it might be argued, from the fall in the share of labour which these figures show for America, that the possibilities of substitution were greater in that country in the later nineteenth century. But the fall is almost certainly to be explained by the decline in the number of self-employed persons as the result of the relative increase of industry and the increase within industry of factory production.

[2] See J. Sawyer, 'The Social Basis of the American System of Manufacturing', *Journal of Economic History*, 1954, pp. 361–79; C. L. Sandford, 'The Intellectual Origins and New-Worldliness of American Industry', *Journal of Economic History*, March 1958. One influence which would repay investigation is the difference between American and British patent law. The expenses of a patent in Britain were much higher—£120 as against about £11 or £12 in America. In America, a patent was granted as a reward to the inventor; in the U.K. as a reward to any person, whether the inventor or not, for bringing the invention to the attention of the public. Thus an American could not get an American patent for an invention of foreign origin: imported inventions could be patented only if the foreign inventor had lived in the U.S.A. for two years. (P.P. 1829, III, pp. 431, 573.)

entrepreneurs were, in general, exceptionally enterprising, and American social attitudes unusually favourable to the exercise of entrepreneurial abilities. Contemporaries frequently commented on the fluidity and receptiveness of American society, on the infectious spirit of improvement, on the tremendous enthusiasm for technical advance, not least as a national, or regional vindication *vis-à-vis* the British. The form of this argument which is least dependent on economic factors is the belief that immigrants contain a higher than average proportion of enterprising people. 'A new settlement', wrote Bagehot, 'voluntarily formed . . . is sure to have in it much more than the average proportion of active men, and much less than the ordinary proportion of inactive'. 'Restlessness of character', wrote de Tocqueville on his visit in 1831, 'seems to me to be one of the distinctive traits of this people. The American is devoured by the longing to make his fortune; it is the unique passion of his life; he has . . . no inveterate habits, no spirit of routine; he is the daily witness of the swiftest changes of fortune, and is less afraid than any other inhabitant of the globe to risk what he has gained in the hope of a better future, for he knows that he can without trouble create new resources again'. And again: '. . . everybody here wants to grow rich and rise in the world, and there is no one but believes in his power to succeed in that'.[1]

It is not the purpose of this essay to deny that there were differences in entrepreneurial attitudes. Some English manufacturers were inhibited by an aversion to the social consequences of technical change. Samuel Crompton, for example, intended the mule to simplify the labour of the spinner in his home and increase the volume and the quality of his output. The application of steam to gigantic mules of 600–700 spindles was no part of his original plan, and 'no man lamented more the changes thus brought about in the life of the people'.[2] William Radcliffe showed a similar attachment to the older forms of industry. His 'New System' was not to be a Factory System. 'Though at that time I was obliged to bring such a number of boys and girls into the factory to work the looms, yet, when the hands had learned to work them it was my

[1] de Tocqueville, *op. cit.* pp. 182, 69.

[2] H. C. Cameron, *Samuel Crompton* (1951), p. 59.

intention to disperse all the looms into the cottages of the weavers throughout the country'.[1]

The mills and village at New Lanark were, it is true, a prototype of the Waltham plan and later developments in Massachusetts—Robert Owen had 2,000 employees under a unified management—but he too was not wholehearted. He admired the landed gentry as against 'the money-making and money-seeking aristocracy of modern times'. Capital was not ploughed back but spent on community projects. In 1821 at the height of his manufacturing career he withdrew from business and sank his money in communist experiments.[2] Among the leading English textile leaders, only Arkwright seems to have been prepared to concentrate actively on increasing output. The attitude of English entrepreneurs was in part a reflection of a general distaste for mechanisation to be found in English society at large, which, in a very extreme form, was expressed by Lord John Manners when he asserted that the time had passed 'when the fallacy could be maintained that the plough, as a productive engine, can compete with the spade'.[3]

The Massachusetts manufacturers had fewer psychological limitations on the urge to expand output. The plans of the leading figures in the industry—Lowell, Tracy Jackson, Samuel Batchelder—were all of a highly ambitious kind. Francis Lowell's plan for a large integrated concern was conceived before the nature of the power supply necessitated it; he foresaw the revolutionary effect of the new techniques and as early as 1815 predicted that cotton cloth, then 33 cents a yard, would some day sell at 4 cents.[4] But perhaps the contrast is most clearly illustrated by the immigrant Samuel Slater. Though he had a revolutionary impact upon American industry, his managerial techniques he brought from England, and he retained them in his mills until the end of his career: 'the family mill', the employment of children and payment in truck. He retained them long after the 'Waltham Plan' had

[1] W. Radcliffe, *Origins of the New System of Manufacture* (1828), p. 30. Radcliffe was often moved by non-economic motives, for example he refused in 1794 to sell yarns to foreigners; 'although at the time I should have been glad of an order from any English or Scotch manufacturer, at a lower price than they (the foreigners) offered to give me, I declined their offer of purchasing yarn from me' (p. 11).

[2] The *Life of Robert Owen by Himself*, with introduction by M. Beer (1920), p. vii.

[3] *Young England Addresses*, delivered 1844, published 1885.

[4] C. Cowley, *A Business Guide to Lowell*, p. 64.

proved its superiority as a managerial system. He never admitted the power-loom into his organisation and turned down the suggestion of his partners that they should do so, with the assertion that, since they had made their fortune in spinning, that business alone was enough for them.[1]

Thus the ingrained traditionalism of the English, while not restricting the growth of the industry as a whole, impeded integration and growth within individual firms. In the U.S.A. the entrepreneurial force which gave rise to a firm was less likely to be halted by conservatism, by unwillingness to accept the social consequences of rapid growth, and by the attractions of other callings. 'The American', it is said, 'is more eager to push on to big things than the typical Englishman'.[2]

There is no doubt therefore that there were differences in entrepreneurial attitudes, and in the climate of opinion in which the entrepreneurs operated. The question at issue is how far these attitudes were simply a reflection of economic circumstances and how far they had an independent existence. As we have suggested economic circumstances would explain some difference in the behaviour of American and English entrepreneurs: where there were fewer alternative uses for his funds, the American manufacturer would be prepared to reinvest his profits more vigorously than his English counterpart.[3] The greater range of alternatives in England, of course, had social roots. They existed because England was a country ruled—at the local and national level—by its landowners, a society where political and social power was so heavily biased towards landownership that until 1832 the principal centres of industry were not represented in Parliament. But there still remains the question how far the English preference was one of rational calculation and how far, on the other hand, it had become ingrained in attitudes and institutions. Would the American

[1] G. S. White, *Memoir of S. Slater* (Philadelphia, 1836), pp. 375ff.

[2] Chapman, *op. cit.* pp. 168–70. Comments to a similar effect are common in the early part of the century. J. S. Buckingham commented on 'the extreme eagerness which the people of this country continually display to "go ahead" as their phrase is, at the utmost possible speed and leave all the rest of the world behind' (J. S. Buckingham, *Eastern and Western States*, I, p. 545). 'The business men of the country are so absorbed in their various pursuits, that any and everything which does not strictly advance these, is thought to be unworthy of their attention'. J. S. Buckingham, *America, Historical, Statistic and Descriptive* (London, 1841), III, p. 169.

[3] See pp. 72–4 above.

manufacturer in England, with the English range of alternatives have disposed his funds in the same way? And the Englishman, faced with the American alternatives, have acted in the same way? How far did attitudes have a life independent of the choices which had given rise to them? No positive answer can be given to this question. The example of Slater does perhaps suggest that there is some residue of 'conservatism' among the English entrepreneurs which cannot be explained by economic circumstances in the strict sense. But, on the other hand, English entrepreneurial attitudes may have been influenced by the abundance of labour. For the social consequences of technological unemployment were serious, and this may well have engendered an attitude of caution among entrepreneurs all the more difficult to remove because it was ingrained and not based upon explicit reasoning. Even Ricardo, for all his belief in Say's Law, became worried about technological unemployment, and Sismondi had many sympathisers even if his economics was disliked. With Robert Owen the problem became an obsession.

An allied possibility to the one just considered is that Americans were, from the circumstances of their life, better at inventing than the English. Americans, it might be argued, were good at inventing machines for the same reason that they were good at inventing constitutions: they lived in a new society, free from the preconceptions and attachments of longer established societies. Benjamin Franklin does not need labour-scarcity to explain him. Among English mechanics, thought James Nasmyth in 1854, 'there is a certain degree of timidity resulting from traditional notions and attachment to old systems'. In America, on the other hand, 'there is not a working boy of average ability in the New England States ... who has not an idea of some mechanical invention or improvement in manufactures, by which, in good times, he hopes to better his position, or rise to fortune and social distinction'.[1] 'English overseers', wrote one American in 1851 with reference to machine-shops, 'are trained too much to one thing or machine and do not *adapt* themselves readily to circumstances—fancy

[1] P.P. 1854, XXXVI, p. 13. 'The American working boy develops rapidly into the skilled artisan ... The restless activity of mind and body—the anxiety to improve his own department of industry—the facts constantly before him of ingenious men who have solved economic and mechnical problems to their own profit and elevation, are all stimulating and encouraging.'

everything wrong which they have not been accustomed to'.[1]

In addition, the absence in America of European traditions of very high craftsmanship may have been a positive advantage in the development of new mechanical methods. In a country where labour was scarce and where there were no long-established institutions for training, the highest forms of craftsmanship were less likely to develop. Thus in America much of the ability of the type which in Europe would have gone into craftsmanship was devoted to ordinary manufacturing operations and had greater opportunity therefore of making improvements to them. Americans of craftsman ability thus devoted themselves to mechanical ingenuity; not being trained to the most intricate invention, they nevertheless turned very successfully to more economically useful gadgets. Moreover, the wide geographical dispersion of industry forced manufacturers in the early days to rely on their own mechanical ingenuity. 'In a new country', wrote one observer, 'he (the skilled workman) becomes almost entirely dependent upon his own resources and ingenuity alone for the supply of those means and materials which, in a more advanced stage of manufacturing progress, the division of labour abundantly aids him in procuring'. The necessity of supply 'has been the means of originating many ingenious machines'.[2]

So far, however, as a capacity for fundamental invention can be assessed, as distinct from the application of inventions, Europeans seem to have been ahead of the Americans. Some of their inventions were more 'automatic' than any American invention: for example the mechanical cannon, designed by the Duke of Orleans in the 1780's, which fired automatically every day at noon; the various machines to play music mechanically; the mechanical writer of P. J. Droz; the Jacquard loom (1801)—an automatic loom for weaving figured silks and controlled by punch cards similar to those used later by Charles Babbage in his mechanical calculator. Babbage's calculator was more intricate and ingenious than anything in America. Even the idea of interchangeable parts was not American but French in origin, and Samuel Bentham and Brunel in England and Bodmer in the Black Forest both approached the system, the former in the manufacture of pulley blocks and the latter in the manufacture of guns.

[1] Gibb, *op. cit.* p. 366.
[2] P.P. 1854, XXXVI, p. 13.

It is probably true that the inventive mind was more likely to apply itself to economic ends in the U.S.A. than in England; Benjamin Franklin turned his intelligence to mechanical gadgets while Jeremy Bentham, with a similar cast of mind, applied his inventiveness to social engineering. If Babbage had been an American he would surely have devoted his mechanical ingenuity to more immediate and utilitarian purposes. But British industry in this period was so much larger than the American that it posed a wider range of problems. At any time there are always many new ideas which spring not from any particular pressure or need but from the observation by intelligent and experienced men that things can be done better; and the more industrialised country provided more opportunities for making such observations. As even a casual reading of Samuel Smiles biographies of the engineers will show, there was no lack of creative ingenuity in England.

New ideas were often the least important part of invention or at least the part which was most easily borrowed. What determined whether and where the ideas bore fruit was the practical capacity brought to bear on them. 'An invention' wrote Bessemer 'must be nursed and tended as a mother nurses her baby or it inevitably perishes'.[1] But here again England was initially very much better endowed with the engineering skills required for such nursing; for this was the great age of English machine-tool invention, the age of Bramah, Maudsley and Nasmyth. And it was precisely on their capacity to turn ideas into workable propositions that the Englishmen of this period prided themselves. 'We have derived' said a leading engineer 'almost as many good inventions from foreigners as we have originated among ourselves. The prevailing talent of the English and Scotch people is to apply new ideas to use and to bring such applications to perfection, but they do not imagine so much as foreigners'. A considerable part of English technical progress had consisted of developing inventions 'that never have risen to any importance in the foreign countries where they were first imagined, because the means of applying inventions abroad are so very inferior to ours. . . . Even after they (the foreigners) got them back in the improved state to which we had brought them, although they received their crude ideas matured, like children sent home after having been

[1] Sir Henry Bessemer, *Autobiography* (1905), p. 50.

educated and become fully grown by boarding for years without expense at a better school than home, still they have not been able to set them to work so extensively and perfectly as we have done'.[1]

The principal difference between America and England was not that the Americans were more inventive or better able to develop their inventions; it lay in the objects to which these abilities were devoted. British inventors in the nineteenth century were not preoccupied with labour-saving, for this was not primarily what the British economy required from its ingenious men. William Fairbairn for example, one of the most intellectually prolific of English engineers devoted his energies to improvements in the performance of textile machinery, the application of iron to building and of steam-power to water-transport, the construction of bridges, and the manufacture of steam engines. These were the problems set by the needs of his customers or by the intellectual interests of his friends. The incidental effect of his inventions was no doubt to save labour and in some cases to save a great deal, but only one of them—the riveting machine—was devised specifically for this purpose.[2] Much the same was true of Henry Bessemer's galaxy of inventions, which were designed principally to improve the quality of products or the efficiency of machinery or to save natural resources. One of his most important inventions—a method of making bronze powder—was, it is true, made in order to save labour and did lead to the creation of self-acting machinery. But the history of this invention shows that exceptional circumstances were necessary to produce in England the desire to save labour which was widespread in America. In this case labour-costs in the hand process were exceptionally high, and therefore the rewards for the invention of a mechanical method correspondingly great. Brass probably costing

[1] P.P. 1829, III, p. 533. It was suggested by one witness before the Committee on the Export of Machinery (P.P. 1841, VII, pp. 111–19) that new ideas for carrying out processes by machinery were devised in the U.S.A., but were subsequently improved and simplified by English artisans who had greater aptitude for this work because more machine-making was carried on in England, and English mechanics were more specialised. 'England has much greater facilities for carrying on the machine-making business than any other country, inasmuch as the great demand which the machine-makers have for the supply of the manufactures of this country enable them to subdivide the different branches of machine making . . .' (pp. 124–5, Qu. 1608).

[2] *The Life of Sir William Fairbairn*, edited and completed by William Pole (1877), pp. 163–4.

no more than sixpence was worked into powder which sold retail at £5. 12*s*. 0*d*. a pound, and it was this fact which prompted Bessemer to reflect that it 'offers a splendid opportunity for any mechanic who can devise a machine capable of producing it simply by power'. Furthermore when Bessemer had invented his new process, in order to keep it secret, he limited his labour force to three relations and a watchman, that is he created conditions of quite exceptionally severe labour-scarcity and this meant that he had to 'design each class of machine to be what is called a self-acting machine'. The machine he produced was more automatic than any American machine at this date, but the circumstances which prompted it were most exceptional; and because Bessemer succeeded in keeping it secret for over thirty years it exerted no influence upon techniques in other industries.[1]

Whitworth's proposals for a uniform system of screw threads and the use of standard gauges are another instance of a lack of concern with labour-saving. Though they certainly saved labour, he did not arrive at them in an attempt to achieve this end, but from his experience of the practical inconvenience of the confusions and delays caused by lack of uniformity. And he commended his proposals, not because they saved labour, but because they saved equipment. 'The immense consumption of bolts and nuts in fitting up and working machinery may give some idea of the extent to which greater accuracy might be productive of economy.'[2]

A number of labour-saving inventions offered to the public by Englishmen in this period failed to find a market. Bessemer developed a type-composing machine which saved a great deal of labour and could be worked by women, 'but it did not find favour with the lords of creation, who strongly objected to such successful competition by female labour, and so the machine eventually died a natural death'.[3] In 1858 Dr William Church

[1] Bessemer, *op. cit.* pp. 54, 61. For the extraordinary range of Bessemer's inventions, see pp. 329–32.

[2] Joseph Whitworth, *On a Uniform System of Screw Threads.* Communicated to the Institute of Civil Engineers, 1841.

[3] Bessemer, *op. cit.* pp. 43–47. It would be easy to multiply the examples of labour-saving devices and methods which were fruitless. For early English experiments with a steam plough and with machinery for making nails, see Pole, *op. cit.* pp. 97, 100. In 1794, Robert Fulton patented an invention for the cheaper and speedier digging of canals, but it was not adopted in England, and in 1797 he went to the U.S.A. (*Life of Robert Owen by Himself*, pp. 90–97). Bessemer invented a

of Birmingham patented a number of machines for the making and graduating of rulers which 'displayed a vast amount of thought and ingenuity' but which were not adopted. 'When it is known', said a commentator in 1866 'that the 540 divisions in the Gunter's line which may be seen on the brass slide of any carpenter's slide rule are cut singly and accurately by hand in ten minutes . . . it will not be surprising that in this instance manual labour should be cheaper and more to be relied on than machinery, designed . . . to draw all those divisions at a single operation'.[1]

Though the idea of interchangeable parts was applied by Brunel and Bentham under pressure of wartime demand, it was not taken up by any other English industry. 'The interchangeable system of manufacture, in a well-developed form, was in operation in England in the manufacture of ships' blocks at Portsmouth shortly after 1800; and yet this block-making machinery had been running for two generations with little or no influence on the general manufacturing of the country, when England, in 1855, imported from America the Enfield gun machinery and adopted what they (the English) themselves styled the "American" interchangeable system of gun making'.[2]

A number of inventions in the textile industry in the nineteenth century were made in Britain but practically applied and developed principally in the U.S.A., for example the principle of the warp-stop motion, invented in England in the 1820's; the idea of automatically renewing the supply of weft, which was hit upon by William Rossetter about 1860; the electrical warp-stop motion devised by a Bradford man.[3] The Northrop loom was invented by James Northrop of Keighley but he migrated to America because English machine manufacturers showed no interest in his invention. The first practic-

machine for sawing plumbago for lead pencils, a difficult art to acquire and one which had hitherto defected all attempts to replace hand-labour by machinery; his process saved raw materials as well as labour, but the firm to which he offered it rebuffed him. (*Autobiography*, pp. 34–6).

[1] *The Resources, Products and Industrial History of Birmingham*, ed. S. Timmins (British Association Handbook, 1866), p. 631.

[2] Roe, *op. cit.* pp. 5, 139. See Whitworth's comments on the failure to extend Bentham's woodworking machinery in England (P.P. 1854, xxxv, p. 116).

[3] T. M. Young, *op. cit.* p. 136. Side drawing, another idea of English origin, was developed in the U.S.A. not in England (Cole, *op. cit.* I, p. 354). There were, no doubt, cases of the reverse process (Cole, *op. cit.* I, pp. 355–6) but they were less common.

able continuous mill for the rolling of iron was patented in England by James While and George Bedson; Bedson's mill was exhibited at the Paris Exhibition of 1867 and introduced into America the following year and adopted more widely there than in the country of its invention.[1]

Shoemaking machinery was invented first in England, failed to be adopted and had later to be introduced from America;[2] the early improvements in woodworking-machines were made in England, but between 1815 and 1851 all the progress in this field was in the U.S.A. and the English industry in the 1850's and '60's had to be modernised with American ideas. The first improvements in signalling were made in England: block signalling began in 1846 and a patent for an interlocking signal system was taken out in 1856. The English interlocking machine was imported into the U.S.A. in 1876, and in 1880 Westinghouse acquired the American rights for interlocking switches and signals, and thereafter the technical development of signalling systems was American.[3]

Other inventions were made almost simultaneously in both countries but subsequently developed more rapidly and extensively in the U.S.A., for example, the condensing process in wool, worked out by Goulding in Massachusetts before 1832 and patented in England in 1834.[4] A series of patents for screw-making machinery was taken out in England, but they

[1] D. L. Burn, *The Economic History of Steelmaking* (Cambridge, 1940), p. 193. Abraham S. Hewitt, *op. cit.* p. 25.

[2] Brunel invented machines for the making of shoes. C. Babbage, *The Economy of Machinery and Manufactures* (London, 1832), p. 14. There is an illuminating case of the role of the two countries in developing inventions in the S.C. on Patent Laws. Two men, Sharp and Whittemore, patented an invention about 1799 for a self-acting machine for making wire cards for preparing wool and cotton. 'The machine was very ingenious and was made to operate with rapidity; but the cards produced by it were too coarse and imperfect to be used with advantage in this country (that is England) where the art of card-making by hand had previously been brought to great perfection: the inventor therefore carried his machine to America where coarse cards were more in requisition; and as our laws prohibited the exportation of any of our cards the want of efficient card-makers in America rendered a self-acting machine of value there, although not very perfect in its operation; he suceeded in America so far as to carry on a trade, and by practice improved his machinery, till it supplied the American demand very well.' An American merchant then re-exported the invention to England, took out an English patent for the improved machine in 1812, and, after making further improvements, established—the witness asserted—a considerable trade. [P.P. 1829, III, p. 548.]

[3] H. G. Prout, *A Life of George Westinghouse* (1922), p. 210, 214.

[4] Cole, *op. cit.* I, 105.

were not adopted on a scale which enabled English manufacturers to perfect them technically, and the manufacture of screws was not revolutionised in England until the introduction from the U.S.A. of self-acting machines.[1]

In America a higher proportion of ability was devoted to the invention of devices designed to save labour; and of the new ideas thrown up by chance the Americans were more likely than the English to select for practical application and further development those which saved labour. In so doing they strengthened the mechanical abilities favourable to the solution of labour-saving problems. We shall enquire later whether this preoccupation with saving labour was more fruitful than a preoccupation with saving some other factor of production. At the moment the argument is only that the success of Americans in the field of labour-saving invention was not the result of some innate superior ingenuity, but of 'the peculiar value of labour-saving machinery to a nation of moderate numbers dwelling in a country of redundant soil'.[2] Where the problem was not principally one of saving labour and the solution did not lie either in labour management or mechanical engineering, the Americans were not in fact conspicuously successful. In the production of iron and steel, for example, the American Commissioner to the Paris Exposition of 1867 reported that in quality of product and invention the British as well as the French and Germans were far superior to the Americans.[3]

To a considerable extent American inventive ingenuity was a result rather than a cause of mechanisation. The American factor-endowment stimulated mechanisation, and the mechanisation gave scope for the development of mechanical skills. In the same way the availability of trained specialists later in the century is to be regarded principally as a consequence of mechanisation. The more capital-intensive methods depended for their proper functioning upon trained specialists—administrative as well as engineering—and any influence which increased the supply of such specialists clearly facilitated the adoption of such methods. The supply of specialists is, of course, influenced by autonomous developments in the educational system, but if the U.S. industry made more use of them in the

[1] Timmins, *op. cit.* p. 607.

[2] Tenche Cox, *State Papers Finance*, II, p. 676.

[3] Abraham S. Hewitt, *op. cit.* p. 19.

later nineteenth century this was principally *because* American industry was more highly mechanised.

Another influence on the equipment of American industry was the nature of the demand facing American manufacturers. American demand was for uniform standardised types of product. The demand facing English manufacturers was more variegated simply because they served a much greater diversity of markets.[1] The fact that products were standardised made it easier to apply machinery to their production. The American demand was not only more homogeneous; it was also predominantly a demand for relatively low-quality goods.[2] And this too was favourable to mechanisation. For many of the new methods in textiles, for example, when first introduced were too clumsy to be suitable for anything except coarse products. They were therefore more likely to be adopted by manufacturers who concentrated on this type of product. Moreover these manufacturers were then in a better position to adapt the machines to the production of finer yarns and cloths. Thus the delay in the adoption of the power-loom in England has been attributed to the fact that, on its first introduction, it was suitable only for coarse fabrics because only the coarse yarns were strong enough not to break under mechanical handling. The self-acting mule was at first suitable only for spinning coarse yarns. And the same was true of ring-spinning.

But here again it is a question whether the standardised demand was principally cause or effect of mechanisation. The total volume of standardised production in England in the early nineteenth century was greater than in the U.S.A.—in England as a whole and probably in a large number of individual concerns. That a higher proportion of American demand was for standardised goods was itself the result—direct and indirect—of the country's factor-endowment. The high price of labour,

[1] 'The immense variety of products in each branch of the trade makes it very difficult to standardise patterns or to adapt automatic machinery. A single forge may, for instance, have to deal with as many as 2,000 different patterns of pocket-knife blades' (G. I. H. Lloyd, *The Cutlery Trades* (1913), pp. 55, 208, 387).

[2] The American consumer would not pay extra for quality: 'The material being expected to last for a single season, is purchased of a quality to do that, and no more. The next season the customer supplies himself again.' 'This habit of almost constant change is said to run through almost every class of society, and has . . . a great influence upon the character of goods generally in demand which . . . are made more for appearance, and less for actual wear and use, than similar goods are in England' (P.P. 1854, XXXVI, p. 76).

and the abundance of land and the relative equality in its distribution ensured that a higher proportion of income than in England went to the middle-income groups who demanded the type of goods which lent themselves to mechanisation; and the cheapness of food ensured a high *per capita* demand for industrial goods. Indirectly mechanisation ironed out much of the original diversities of demand. In the eighteenth century, as Professor Wertenbaker has shown, there was a great variety of types of product; but these specialities, most of them national specialities—German pottery, etc.—were driven out because, by their very nature, they were labour-intensive products and could not sustain price competition of more mechanised standardised processes. High labour-costs could be offset by mechanisation most easily where the products were plain and uniform. Therefore these products fell most in price; therefore people bought them in preference to the goods of variety and distinctive quality.

The character of American demand was also related to its rate of growth. A large part of the demand facing American factory industry was new demand, as opposed to demand diverted from industry organised on the putting-out system. Being new in this sense, there was no reason why it should not be homogeneous—it was a demand for what mechanised industry was producing. By contrast a large part of the demand for the products facing English factories was diverted from other suppliers, that is it was a demand with idiosyncracies already formed.

Another influence which is independent of factor-endowment is the timing of American industrial development, which also has some bearing on structure. This is particularly relevant to the mechanisation of the Massachusetts cotton-textile industry. The English industry developed earlier: heavy investment in mechanical spinning came *before* mechanised weaving was technically possible; that is the cotton-textile industry in England grew gradually, and the speed of growth and the forms it assumed were influenced by the pace of invention. The fundamental textile inventions extended over a long period, say, from the successful adoption of Hargreaves' jenny in the late 1760's to the adoption of the self-acting mule in the 1840's and '50's. The slow rate of invention itself set up social reactions which would not have been present had there been a sharp technological break; each fresh technical development created vested interests, among capitalists as well as labourers, in particular forms of industrial organisation and production which acted as an impediment to the adoption of succeeding technological developments. In contrast, the New England cotton textile industry developed later at a time when the pace of invention was much less of a limitation. It took over from England the basic inventions which had already reached a reasonable degree of technical efficiency. For these reasons there was no reason why the newly-established concerns in New England should not be large integrated concerns. The fact that the Massachusetts mills were larger than the English, though it owed something to the ease of incorporation in America and to the nature of the water-power available, was also partly the result of the late start of the industry; and because the mills were larger it was easier for them to accept innovations without radically changing their technical structure; the size of the unit was large enough to overcome the effects of indivisibilities.

In the case of cotton textiles there was also a fortuitous element. The form of the Massachusetts industry was greatly influenced by the success of the Boston Manufacturing Company founded at Waltham by Francis Lowell in 1813—the first large integrated concern employing power in weaving as well as spinning. The success of the Waltham plan was no accident, but the plan itself has a number of accidental elements about it. It was partly the result of the coincidence of a boom in textiles with a depression in foreign trade. Since the outlets in foreign trade were temporarily restricted, Lowell turned to textiles, and since he was a wealthy successful merchant, not restrained by limited ideas about size, he built on a large scale. Once the decision to build a large plant had been taken, the scarcity of labour made it necessary to establish a large community settlement to attract labour, and the heavy investment this entailed became an additional reason for making the best use of the labour force. Once manufacture on the Waltham plan had been shown to be a success, this type of community settlement became an independent influence in favour of units of manufacture sufficiently large to return a reasonable profit on the total outlays on site development. But this type of settlement was only necessary after it had been decided to proceed on a large scale rather than on the 'family plan' on which the Rhode Island mills operated.

It is not the purpose of this essay to argue that the industrial development of America in the first half of the nineteenth century can be explained by reference exclusively to 'scarcity of labour'. But, in so far as any simple formula can, this explains a great deal, including a number of features in the American economy which are sometimes presented as independent influences on industrial growth.

IMMIGRATION AND TECHNOLOGY

With the passage of time changes occurred in the conditions of labour-supply in America. The cost of occupying new land rose in relation to the earnings of industrial labour, and in the 1850's it was less possible than in preceding decades to meet the costs of farm-making on the frontier by the labour and sacrifice of the farmer and his family.[1] The start of heavy immigration in the 1840's and '50's narrowed the disparity between the cost of labour, in terms of output, in the U.S.A. and in England. Since immigration was greatest in American booms, it also reduced the disparity between the inelasticity of labour-supply between the two countries; during periods of high investment in America the supply of labour was less inelastic than it would otherwise have been, or than it was in earlier decades, and since American and English investment booms had some tendency to coincide, British labour-supplies became less elastic than they had been earlier.

It might be supposed that these changes would reduce the significance of the effect of labour-supplies on technology. On the other hand, American labour was still dearer than the English. Productivity in American agriculture was rising sharply from the 1850's, and the building of the railways made land on the frontier more physically accessible than otherwise it would have been. For many parts of America it was still true that it was the yield of labour in agriculture, including the bringing into cultivation of new land, which made industrial labour expensive. This was the explanation which seemed natural to the foremost American ironmaster of the later 1860's. The ironmaster, Abraham Hewitt, who was associated

[1] C. H. Danhof, *op. cit.* p. 318.

with works in New York State, in New Jersey and in Pennsylvania, was asked by the Commission on Trade Unions in 1867 why wages were so much higher in the U.S.A. than in Britain, and his reply is worth quoting. 'We have a new country of immense extent, a very fertile soil, sparsely populated even in the populated parts, and to a very large extent entirely unoccupied. Every enterprising man can, even near the centres of population, purchase land upon credit if he has not any capital, or he may go west and have land for nothing (that is under the Homestead Act). . . . Of course the rate of wages is regulated substantially in our country by the profits which a man can get out of the soil which has cost him little or nothing except the labour which he himself or his family have put upon it. This is the element which in my judgement determines the standard of wages in the United States which an ordinary labourer will derive, from the fact that a man who thinks he is not getting enough takes to the land. . . . In other words what you call rent to some extent enters into the question of the value of labour in our country. You add to what would be the ordinary value of labour in this country (that is England) the element of land, and you arrive nearly at the value of a day's labour in America. Rent does not exist in America, or rent upon uncultivated lands exists to a very slight extent'.[1] To attribute the dearness of American labour to its high returns in agriculture was attractive as a reason for protection which did not reflect badly upon American industry. But Hewitt was not a protectionist and there is no reason to believe that he was not essentially recording his own experience.

Moreover to an increasing extent the role which agriculture had played with respect to industry as a whole was now taken over by those industries where the search for labour-saving methods had more than compensated for the high cost of American labour; in varying degrees in different places, these were setting the pace for those industries which had been less successful in achieving this end. The expansion of demand for labour by industries where it was very productive put pressure on the labour supplies of those industries where labour productivity remained lower.

American labour therefore continued to be dear labour by English standards, and after the beginning of heavy immigra-

[1] R.C. on Trade Unions, P.P. 1867, xxxii, p. 175.

tion it was the dearness of American labour rather than the unresponsiveness of its supply which influenced the choice of technique. Even so, difficulties about obtaining additional labour were still more apt to arise in America than in England. A period of exceptional scarcity occurred during the Civil War which absorbed 60 per cent of the man-power of the northern states and undoubtedly accelerated the mechanisation of both industry and agriculture.[1] This exceptional episode revived in an accentuated form the scarcity of labour of earlier times. Moreover, though after the beginning of heavy immigration the labour-supply in America increased more rapidly for a given rate of investment than it had done earlier, the rate of investment was more rapid in the decades after the civil war than it had been before.

Despite immigration, therefore, labour conditions gave American investment a bias towards labour-saving, capital-intensive techniques even in the second half of the nineteenth century. It has also been suggested that the *character* of the immigration was directly favourable to mechanisation. Some writers argue that it was the availability of cheap, unskilled immigrant labour, combined with the scarcity and belligerency of the available skilled labour, that promoted mechanisation. 'To the American employer', writes Dr Erickson, 'the scarcity of skilled labour, his unwillingness to train it, the difficulties of recruiting it abroad, and the unsatisfactory result in trade unions when he did, combined to recommend a policy of increasing mechanisation to free him from the demands of the skilled European workmen'.[2] An analogous line of argument is put forward by Professor Brinley Thomas: 'The introduction of automatic machines eliminating human skill over a wide field was stimulated by the incursion of such a considerable volume of cheap labour: it was profitable to invest in capital-saving equipment'.[3] Heavy immigration, he argues, put a premium on processes which replaced skilled labour by machines using a relatively large quantity of low-grade labour.

A scarcity of skilled labour affected the choice of techniques

[1] For the effect of the Civil War on the labour supplies of a Massachusetts machine works see Navin, *op. cit.* pp. 54–5. There are some interesting observations in the Commission on Trade Unions, pp. 179—80.

[2] C. Erickson, *op. cit.* p. 124.

[3] B. Thomas, *Migration and Economic Growth* (Cambridge, 1954), pp. 158, 165. See also Handlin, *op. cit.* p. 62.

in contrary ways.[1] Since the production of more complicated machines required more skill than the production of simple machines, a high premium on skill would tell *against* the adoption of the more complicated machines, other things being equal. But, it may be argued, while the more complicated machines compared with the simpler machines required more skill to make them, they may have required less skill to operate them. The same analysis as we have applied to techniques, classified according to their employment of labour and capital, could, in principal, be applied to techniques classified according to the extent to which they employed various grades of labour. Where there was a shortage of skilled as compared with unskilled labour, it might pay the industrialist to install a more unskilled-labour-intensive technique, that is one which increased the productivity of his skilled labour even at the expense of some increase in his unskilled-labour-costs per unit of output. The unbalance in the labour force would provide an incentive to exhaust the skilled-labour-saving possibilities of existing techniques and to invent new techniques for the purpose.

The force of this incentive to technical progress would depend on the relative responsiveness to demand of the various grades of labour. Where grades are equally responsive, or equally unresponsive, or are easily substituted for each other, the character of labour supply provides no incentive to adopt and develop techniques so as to employ a higher proportion of less skilled labour. Nor are there likely to be marked differences in responsiveness in a mature industrial economy in which the chain of substitutes in labour is continuous. In these cases an increase in demand for skilled labour is more likely to lead to slight reshuffles of the labour on particular operations and to the training of skill than to innovations designed to employ unskilled at the expense of skilled labour. It is where there is a marked and persistent difference in responsiveness that the incentive to do this exists. And it can be argued that this was the case in the U.S.A. once large-scale immigration had started and before the American labour-supply had become fully adapted to industry. In the 1850's and '60's additional skilled labour had, generally speaking, to be recruited from among the limited stock of native-born, while additional general labour

[1] See the discussion on pp. 151–6 below.

could be obtained from immigrants who were being eased out of Europe by agricultural changes and population pressure. And while in the earlier decades of the century cheap land had exerted its pull mainly on general labour and the premium on skill had probably been lower than in England, the rise in the effective costs of settlement in the 1860's and the initial attraction of the towns for immigrants meant that the pull of the land now exerted relatively more influence on the skilled. The lowest stratum of industrial labour, said an American industrialist in the later 1860's, was confined to Irish and Germans and contained only a very small proportion indeed of Americans: 'if an American wants to labour with his hands, or in an occupation not requiring mechanical skill, he takes to the land as a general rule, or a farm . . .'.[1] And the same employer thought that the premium on skill was higher in America than in England. Furthermore, the supply of native skilled labour was often controlled by union action, while immigration made it difficult to unionise unskilled labour. One might argue that in these circumstances, though the greater abundance of labour in general reduced the incentive to devise methods of replacing labour by machines, the unbalance of the labour supply gave a strong inducement to devise ways of replacing skilled by unskilled labour, so that the course of invention was turned to take advantage of the types of labour that were responsive to demand at the expense of those that were not.

How far the effect of labour-scarcity in general was, after the mid-century, complicated by the much greater scarcity of skilled than unskilled labour it would require a very detailed examination of technology to establish. And perhaps the point never can be established, since it is difficult to distinguish in practice between the effect of a technique on labour and capital inputs and its effect on the inputs of various grades of labour. It may be doubted whether there were many new methods which depended for their effect solely on replacing skilled by unskilled labour: it was usually the reduction of *all* grades of labour per unit of output that was important rather than the change in the grades. In most cases, I should guess, the new automatic methods would still have been the most profitable even had unskilled labour been less abundant, that is, the increased availability of cheap immigrant labour did not

[1] P.P. 1867, XXXII, Qu. 3822.

change the choice of the industrialist from a less to a more highly mechanised method.

The principal effect of immigration was not on the manufacturer's choice of technique but on his ability to give effect to his choice. In the first place, immigrant labour made it somewhat easier to frustrate attempts by skilled workers to prevent the adoption of more highly mechanised methods. In the second place, the increased supply of labour removed a restraint on the rate of investment; it facilitated a large amount of investment at a time when there was a large body of new techniques to be absorbed. Immigration encouraged the American industrialists to buy additional machines, and they naturally bought the most economic machines, but machines which would have been the most economic even in the absence of heavy immigration. Indeed it might be said that the U.S.A. was lucky in this respect. After the mid-century, the labour-supply available to American industry became less inelastic; heavy immigration damped down wage rates; so that profits were high in the U.S.A. and therefore investment more rapid, at a time when technical progress everywhere was considerable. The U.S.A. had her period of most severe labour scarcity in the first half of the century and was stimulated by it to technical progress at a time when autonomous technical progress was inconsiderable; in the second half of the century when the stream of autonomous technical progress was much more vigorous, labour, though still 'scarcer' than in England, was more abundant than it had been early in the century.

V

LABOUR-SUPPLIES AND TECHNOLOGY IN BRITAIN

We shall now examine capital accumulation and labour supplies in England.

In eighteenth-century England, the floor set to the industrial wage by earnings in agriculture was much lower than it was in America; moreover, for a given rate of capital accumulation, the supply of labour was more elastic. The rate of capital accumulation in English industry was, however, higher in England than in America. In the industrial areas it was certainly higher than the rate of growth of labour-supplies, as is evident from the fact that wages rose in Lancashire much earlier than in the southern counties and that agricultural wages were higher in industrial than in non-industrial counties. Goods went out faster than men came in.

It is probable that this disparity between population growth and capital accumulation was greatest from 1730 to 1760 when the growth in the adult population of the country as a whole was modest, the growth of foreign trade perceptible, and the transport improvements which facilitated labour mobility still at an early stage. But the disparity remained considerable even after population started to increase in the middle decades of the century. 'Until 1800 or thereabouts' writes Professor Ashton '. . . the supply of industrial labour was, over short periods, inelastic. For though there were in rural England large numbers of under-employed men, most of them were unskilled and unaccustomed to discipline, and only when special inducements (including the provision of houses) were offered, or when want was acute, could there be any quick transfer from the countryside. It was partly because of this difficulty of recruitment that employment in building and construction tended to vary inversely with that in manufacture. At times when activity in both was high, as in 1764, 1772 and 1792 there were complaints of a shortage of labour . . .'.[1]

[1] T. S. Ashton, *Economic Fluctuations in England, 1700–1800* (Oxford, 1959), pp. 173–4. For pressure on existing supplies of labour in 1792–3, accentuated by the heavy demands of canal-building, see the same work, p. 167.

In eighteenth-century England therefore the increase of industrial investment tended to run up against rising labour-costs, and so manufacturers had an inducement to adopt more capital-intensive techniques where these were available, and to invent labour-saving techniques, for reasons similar to those we have already explored in the case of America. Since, until the end of the century, the range of available techniques was very narrow, so that shifts within it afforded few possibilities of escaping a rise in the cost of labour, the incentive to invention was considerable and perhaps not entirely uncomparable with the incentive provided by scarce labour in the United States in the early nineteenth century; for though America's labour-scarcity was more severe, the possibility of mitigating it within the range of existing techniques was greater in the early nineteenth century than in the eighteenth.

The great inventions of the classic industrial revolution occurred in England partly because English manufacturers—for reasons which were principally political—enjoyed a preferential position in good markets, and partly because, for simple geographical reasons, she was best situated to take advantage of reductions in the cost of transport. The landbound industrial centres of continental Europe served markets where the growth of income was slow, and they sent the bulk of their goods overland. The industrial areas of England had a large market in North America where total income was growing rapidly; they had a large home market and could make greater use of water-transport, which cost less than land-transport and in the later eighteenth century was becoming cheaper. The great inventions therefore are partly the result of the fact that, for these reasons, demand for English manufactures was increasing more rapidly than for those of other areas.

But had the supply of English labour been abundant the increase in demand would have been met in England, as on the Continent, simply by expanding the domestic system with equipment of the existing type. In large sectors of English industry this was in fact the response, for example among the iron manufacturers of the Black Country, sometimes because of a local abundance of labour and sometimes because the technical problems of devising means of replacing labour by machines proved intractable. But in industries situated, by historical accident or because of raw materials and power, in the more sparsely-populated parts of England, and employing techniques which lent themselves to improvement, the increase in demand noticeably stimulated labour-saving inventions.

That labour-scarcity stimulated invention in the English cotton-textile industry is an old story. The high rate of investment in this industry showed up the bottle-necks, and particularly the bottle-necks in certain types of labour; and the preliminary inventions of the 1730's and 1740's were attempts to cope with this situation.[1] The advantage of Wyatt's spinning machine (1733) was that it would reduce by one-third the labour for spinning. Ure attributed the lengthening of mules, the invention of the self-acting mule and some of the early improvements in calico printing to the need to economise labour.[2] Indeed the traditional explanation of the rapidity of technical progress in this industry—the lack of balance of its various sections—would be implausible without a shortage of labour. The emergence of the imbalance is not necessarily a sign of general shortage, but only of a deeply-rooted division of labour between the heads of households, who did the weaving, and their dependents, who spun; but with abundant labour, the imbalance could have been corrected, without technological improvements, by changes in the rate of recruitment as between spinning and weaving.

But shortage of labour had repercussions on technology in many more fields than textiles. It perhaps accounts for the fact that England had a reputation as the home of labour-saving inventions: 'few countries' wrote Dean Tucker 'are equal, perhaps none excel the English in the Number and Contrivance of their Machines to abridge Labour'.[3] A pamphlet in 1780 observed that: 'Nottingham, Leicester, Birmingham, Sheffield, etc. must long ago have given up all hopes of foreign commerce had they not been contracting the advancing price of manual labour by adopting every ingenious improvement the human mind could invent'. In agriculture dear labour in the war period promoted the adoption of the threshing machine which

[1] The connection between the shortage of hands and the invention and adoption of machinery is discussed by L. C. A. Knowles, *The Industrial and Commercial Revolutions in Great Britain during the Nineteenth Century* (2nd ed., 1922), pp. 31–4.

[2] Ure, *Philosophy of Manufactures* (1835), pp. 367, 369.

[3] J. Tucker, *Instructions for Travellers* (1757) in a *Selection from his Economic and Political Writings* (New York, 1931), ed. R. L. Schuyler, p. 241.

had been made a success in sparsely populated Scotland in the decade before the war.[1]

In eighteenth-century England, therefore, though the labour available to industry was cheaper than in America and its supply more elastic for a given rate of expansion, it was none the less a sufficient restraint on capital accumulation to divert this accumulation towards the more capital-intensive methods and to stimulate invention. To this extent the contrast between the U.S.A. and England is not simply between a labour-scarce and a labour-abundant economy, but between economies which had their labour-scarcity at different periods or at different phases in their industrial development. Though in all the relevant senses labour was always scarcer in America than in England, England had her period of labour-scarcity while the techniques of the classical Industrial Revolution were being invented; America had hers when these techniques were being widely adopted.

Towards the end of the eighteenth century there were signs that labour-supplies in England were becoming more responsive. It is true that investment was speeding up in the 1780's and '90's, but so was population growth; and the extensive system of outdoor relief operated not only as a system of family endowment to increase the births of potential wage-earners, but also as a subsidy to the wages of adult workers: 'a large portion of the labourers of England . . . received not strictly speaking wages, regulated by the value of their labour, but rations proportioned to their supposed wants'.[2] From the demand side, the fact that many of the inventions were labour-saving relieved the pressure on labour-supplies.

For some time the heavy drain of men for the army and navy during the wars of 1792–1815 offset the forces making for more abundant labour-supplies.[3] There was also considerable demand

[1] N. Gash, 'Rural Unemployment', *Econ. Hist. Rev.*, VI (1935), p. 92. J. H. Clapham, *Economic History of Modern Britain* (1926), I, p. 139.

[2] Nassau Senior, *Industrial Efficiency and Social Economy*, ed. S. Levy (1928), II, p. 260.

[3] See Robert Owen's opinion on the effect of the war. 'The want of hands and materials created a demand for and gave great encouragement to new mechanical inventions and chemical discoveries, to supersede manual labour in supplying the materials required for warlike purposes . . .' (*The Life of Robert Owen*, *op. cit.* p. 171). See also J. Lowe, *The Present State of England* (1822), p. 45, Appendix pp. 12, 13.

for labour for enclosures and in certain years for canal-building. Despite the fairly rapid rise in population, therefore, considerable local shortages and probably some degree of general shortage persisted. This is suggested by the course of agricultural wages. The wages of agricultural labourers in the south of England appear to have risen very little until the war years, and then they rose steeply. To a large extent, of course, this rise merely reflected the rise in the price of provisions, and the occurrence of some years of exceptionally high grain prices, years 'when the clamour (for higher wages) is too loud and the necessity too apparent to be resisted'.[1] But without heavy demands on labour made by Government, directly and indirectly, wages would not have risen so rapidly. There are also specific instances which suggest labour-scarcity in industry, for example the meeting of merchants in Lancashire to devise improvements in the power-loom in view of the shortage of weavers.

After 1815, however, and probably quite abruptly, the force of labour-scarcity as a restraint was very considerably diminished in England, partly because of a diminution in certain demands, partly because of a more abundant supply. The supply of labour was increasing from the following five sources:

(1) After 1815 a large number of men, some 400,000, were demobilised, and a substantial number must have been released from agricultural investment (especially enclosures) which was low from 1815 to the 1850's. 'This surplus' [that is of agricultural labour] said a Sussex farmer in 1837 'has been growing for years; but it commenced with the peace'.[2] The agricultural surplus in the southern counties was principally due to large families and a low level of agricultural investment. 'The competition for work among the labourers is so great, that they underbid each other, and leave the bargain entirely at the discretions of the master'.
(2) Population was increasing very rapidly.
(3) Capitalism was invading the labour reserves of domestic industry. Many of the textile factories, until well into the nineteenth century, drew their marginal labour from rural industry. The villages to the north and north-east of Coventry, for example, in

[1] T. R. Malthus, *Essay on Population*, 1798 (Royal Economic Society Reprint, 1926), p. 35.
[2] Quoted by A. Redford, *Labour Migration in England, 1800–1850* (Manchester 1926), p. 70.

the 1830's and '40's provided a reservoir of country weavers for the Coventry industry.

(4) There was a large influx of labour from Ireland.

(5) Technical changes in agriculture increased the supply of labour available to industry. Though it is true that landless labourers were so numerous in English villages before the main phase of the enclosure movement that the relative increase due to enclosure was slight, enclosure did make *some* difference; by reducing rights of common it tended to weaken the ties which bound the cottagers to the villages and so made them less reluctant to respond to the demand of the factories for labour.

But the main significance of the agricultural improvements was that they prevented accumulation of capital being impeded by shortage of land. This was the restraint on accumulation which Ricardo feared. He did not doubt that an increase in population would be forthcoming when accumulation required it, but he feared that the increase in population could be fed only from inferior land so that, though the content of the labourer's subsistence diet remained the same, its cost in terms of output would rise, and the capitalist would have to pay more in order to obtain labour. From the point of view of the individual manufacturer the effect of a given rise in wages, in terms of output, was the same whether it arose, as in the United States, from rising productivity of labour in agriculture, or, as it threatened to do in England, from diminishing returns to land. But from the point of view of the economy as a whole the effect was different. In America the high productivity of labour in agriculture stimulated labour-saving improvements in agriculture, which raised the productivity of labour; since labour was scarce, it was able to enforce its claim to a share of its increased productivity, and therefore the characteristic improvements in American agriculture increased the wage industry had to offer to attract labour. In England, on the other hand, the scarcity of land stimulated land-saving improvements —enclosure, new rotations, and better drainage. These probably had the incidental effect of saving some labour per unit of output, but not a great deal and, in any case, since there was a surplus of labour in agriculture, labour did not benefit from such increase in its productivity as did occur. The general effect of the English agricultural improvements therefore was to keep down the cost, in terms of output, of the subsistence wage, to prevent any deterioration in the terms which English manufacturers had to make with labour—both existing labour and that which they needed to attract out of agriculture. Until 1815 the offset may have been partial. A considerable area was enclosed during the war years but enclosure, though it was usually a necessary condition of better crop rotations and drainage, did not ensure that these improvements were adopted; conservatism and lack of know-how delayed the adoption of better rotations, and the major improvements in drainage had to await the technical inventions of the 1820's and '30's. It is possible that the restraint which Ricardo feared did in fact operate to some extent before 1815; agricultural prices certainly rose more than the prices of manufactured products, and probably to a greater extent than was warranted by the changes in the relative costs of agriculture and industry.[1] But certainly for the period of thirty or forty years after 1815 there was no restraint on industrial accumulation from the side of agriculture.

Quite apart from the nature of agricultural improvement in England, English agriculture was depressed for most of the period between 1815 and the 1840's because of the heavy agricultural investment undertaken during the wars against France. Because of technical improvements, particularly in draining, it was still possible for farmers in certain areas or in certain types of agriculture to expand output, and because of unresponsiveness to prices, or sometimes the perverse response, the farmers in the areas which were under pressure did not make the appropriate contractions of output. Because agriculture was depressed, industry could obtain additional labour from the reserve army of agricultural labourers on favourable terms; and because food was cheap, the wages, in terms of output, of the industrial workers were low.

For all these reasons there was considerable underemployment of labour in England in the decades after 1815, particularly in the English countryside. There is direct evidence of this in the figures of able-bodied paupers, and also in the course of real wages. Whether there was an increase in real wages down to the 1840's is still in dispute but even on the more optimistic view it seems that any rise was not enjoyed by the great mass

[1] See the discussion in W. D. Cramp, *The Manchester School of Economics* (Stancrd, 1960) pp. 29–33, and M. Blaug, *Ricardian Economics, a Historical Study* (New Haven, 1958).

of unskilled workers. The inconclusive nature of the current debate about living standards in this period is perhaps a warrant for supposing that a substantial and general and demonstrable rise in the real wages of industrial workers did not occur until the 1850's and '60's; it was not until about 1870 that real wages in agriculture began to rise and a steady rise was apparent only in the 1880's.[1]

From 1815 to the 1850's the English manufacturer, in comparison with the American manufacturer, had a very elastic supply of labour at the ruling wage. It need not be supposed that this was a physiological minimum wage; it included an addition for superior diligence and skill and the transport costs of movement into industry; and the employers' fear of social disorder as well as social conscience might maintain it at a higher level than strictly economic forces by themselves would have done. Moreover long-term abundance of labour was compatible with complaints of shortages in particular industries and localities, with a favourable bargaining position for particular groups of workers, and even, on occasion, with a more general sense of shortage at the height of a boom. But in general the shortages were local and temporary.

There were complaints of scarcity in the mid-'30's but these were the result of exceptional circumstances; they were a response to the Factory Act of 1833 which happened to occur on the eve of a substantial increase in capacity in the textile industry. When the factory inspector, Horner, examined the matter, he agreed that occasionally there was an insufficient supply of children especially in country situations, 'either where manufacturers have increased in a more rapid ratio than the infant population, or where improved machinery requiring an additional number of children instead of being worked by adults, has been introduced into a thinly populated district'. He concluded, however, that in general, the complaints were not well-founded since 'in places where the want of children is most confidently stated, numerous new factories are building, and persons are therefore embarking large capitals without having apparently any doubt of finding children to work their machinery'.[2] In the 1830's despite factory legislation and a very substantial increase in industrial output, real wages were

[1] G. H. Wood, 'Real Wages and the Standard of Comfort since 1850' *S.J.* 1909.
[2] Report of the Inspectors of Factories, P.P. 1837, XXXI, pp. 98, 60.

comparatively stationary and do not show any marked response to changes in industrial activity.[1]

In the boom of the early '50's there are signs of some tightening in labour-supplies. 'In England', wrote Sir Morton Peto in 1856, 'floating Rural Labour has within the last few years been absorbed by Emigration, by the prosperity of our Manufactures and Commerce, and recently by the necessary increase of our army'.[2] This was written after some years of rising activity, but it reflects the operation of longer-term forces which were reducing the size of the reserve army of surplus labour. The beginning of heavy emigration in the 1840's absorbed some of the surplus, and the legal limitation of work in factories had a similar effect. The fact that industrial real wages were rising in the middle decades of the century is perhaps evidence that demand for labour was out-running the expansion of the labour force. But for the first half of the century at least the labour available to industry was clearly cheaper and very much more abundant in England than in America.

The abundant supply of labour in England and the low floor set to the industrial wage by agricultural earnings allowed capital accumulation to proceed without pressure from that side. In many respects the behaviour of the English economy in the first half of the century conformed to the operation of a Ricardian model in which additional food supplied was more readily forthcoming than Ricardo himself supposed. What would have happened had fewer land-saving innovations existed or imports been less available it is difficult to say. Perhaps the employers would have had to offer increasingly higher wages in order to obtain additional labour, their profits would have been eaten into, and capital accumulation would have been slower. But it is conceivable that with labour so abundant, the effect of unresponsive food supplies would have been to reduce living standards rather than profits. In any case the Ricardian fears were not put to the test.

With no more than a gentle brake exerted from the side of labour-supplies, considerable new additions were made to capacity—much larger additions absolutely than in other

[1] R. C. O. Matthews, *A Study in Trade Cycle History* (Cambridge, 1954) pp. 221, 147. I hope to deal, in a separate article, with the effect of English factory legislation on technology.

[2] T. Tooke and W. Newmarch, *A History of Prices, 1792 to 1856*, ed. T. E. Gregory (1928), VI, p. 175.

countries—and this allowed scope for the employment and testing of new methods. It was because of the large amount of investment in the Lancashire textile industry that craftsmen—smiths, joiners, shoemakers, etc.—were attracted into the manufacture of textile machines to which they contributed a large number of technical improvements, most of them small individually but significant in the aggregate. For the same reason specialised textile-machine-making firms emerged, and even firms which specialised in the making of particular parts of the various textile machines and so made much better machines.

From this point of view the abundance of labour in England was favourable to technical development, for by promoting a high rate of investment it afforded Englishmen plentiful opportunities of trying out new methods. On the other hand, it blunted the incentive to invent labour-saving devices; and also the incentive to adopt those which were invented, when the price of saving labour was an increase in capital per unit of output. Abundance of labour favoured accumulation with existing techniques—widening rather than deepening of capital—even though the supply of capital might have permitted a technologically more advanced development.

This was the case in agriculture where the effect of abundant labour was reinforced by the operation of the old Poor-law. 'The great spur to improvement to the employer, the desire to make a given amount of labour more efficient is wanting. A threshing-machine would not cost the wages of one man for a year, and would save the labour of two. But the two men are there and must be employed or relieved, so they are set to work with their flails'.[1]

Abundant labour-supplies were also unfavourable to the adoption of labour-saving devices in industry. 'Mechanical improvements—such as the mule and the power-loom—are of slow introduction' wrote the Commissioners on hand-loom weavers. The adoption of the self-acting mule was slow and of the ring-spindle very much slower. And some apparently promising ideas for mechanical improvements were not followed up at all. The explanation may be that English industrialists were slow to see where their interests lay. But it may be that the immediate economic facts did not warrant a

[1] Nassau Senior, *op. cit.* II, p. 242.

faster rate of adoption. A number of the new methods in textiles were ill-adapted to the production of the finer qualities and this restricted their use. But there seems a strong presumption that, over and above such particular causes, the new mechanical methods which saved labour were often initially expensive in capital and that this mitigated against their adoption in a country of abundant labour. Part of the explanation for the slow adoption in England of the self-actor mule was its unsuitability for fine spinning, but even in the medium and course counts it did not dominate the field until the 1850's. It also had certain technical deficiencies—in particular the winding done on the self-actor was defective. But this was as much a result as a cause of its slow adoption. Its slow adoption must in large measure have been due to the fact that though it saved labour it was expensive in capital. As an observer wrote in 1887, 'for long, on account of the excessive cost, very few firms were able to purchase it, and therefore hand-mules were the rule and self-actors the exception'.[1] Similarly the slow adoption of the ring-spindle in England can partly be attributed to the fact that it is best suited to the coarser counts of yarn, but much of the explanation is that this technique saved labour rather than capital.

Moreover since labour in England was abundant it was hostile to labour-saving inventions. As late as the 1840's the tradition in the neighbourhood of Jethro Tull's farm in Hungerford was that he was 'wicked enough to construct a machine which . . . beat out the corn without manual labour'.[2] In the agrarian risings of 1830 there were widespread attempts to prevent the introduction of threshing-machines. These riots 'occurred just when the battle of machines versus arms was well joined. They succeeded not only in smashing much machinery but in delaying its march'. There were also frequent attempts to destroy new machines in the cotton-textile industry.[3] But direct destruction was probably a less powerful impediment to

[1] Chapman, *op. cit.* pp. 69–79.

[2] Ransome, *op. cit.* p. 5.

[3] Clapham, *op. cit.* p. 140. N. J. Smelzer, *Social Change in the Industrial Revolution* (1959), pp. 227–8, 248–9. In agriculture the operation of the old poor law was unfavourable to mechanisation, for it was a subsidy to wages and, in a sense, a tax on agricultural profits. A witness from Huntingdon before the Select Committee on Labourers' wages (P.P. 1824, VI, p. 431) attributed the decline in the number of threshing machines to 'the quantity of labourers we are obliged to maintain'.

mechanisation than more persistent opposition on the factory floor. Senior believed that the owners of old machinery incited workers' combinations to act against their rivals who installed new machinery, and that a common object of strikes was to induce a manufacturer not to use a given sort of machinery or generally to change his mode of managing his business.[1]

The most effective form of hostility to mechanisation was the resistance of workmen to the reduction of piece rates on the more productive machines, for example the longer mules. In the spinning section of the cotton-textile industry the employers in the 1820's and '30's were faced 'by a resolute effort to stop economies in production'.[2] Thus the nature of his labour-force imposed limitations on the manufacturer's choice of technique, and these limitations were more restrictive than those in the U.S.A. For a higher proportion of the English labour-force in the mid-nineteenth century was accustomed to certain industrial routines. This was principally because a very large part of industrial growth in England took the form of expanding operations in established industrial centres; whereas in the U.S.A. industry was established in new areas. English employers who could draw on local supplies of experienced labour derived some benefit from the fact; but because much of the labour was habituated to certain forms of work and methods of operation, it was apt to resist change. This resistance was not confined to changes which threatened to depress earnings; where a new method involved a considerable disturbance to existing ways, the workers might resist it, even though it promised an improvement in earnings. An American industrialist developing industry in a new area had difficulty in recruiting labour, but he was less limited in his choice of techniques by the stereotyped nature of his labour. Certain areas in England at certain periods, for example new coal- or iron-fields, may have enjoyed similar advantages especially at the height of a boom, and it would be interesting to investigate how far the competitive superiority of for example the South Welsh coal-

[1] Nassau Senior, *Essays Historical and Philosophical* (London, 1865), II, ch. VII. 'Combinations and Strikes', a report written in 1830 and revised 1841.

[2] Chapman, *op. cit.* pp. 76, 78–9. For the attitude of English labour to mechanisation see S. and B. Webb, *Industrial Democracy* (1897), I, ch. VIII; R.C. on Trade Unions, P.P. 1868–9, XXXI, p. 394. For instances in which riot and machine-breaking delayed mechanisation see E. J. Hobsbawm, 'The Machine Breakers', *Past and Present*, vol. I (February 1952), p. 67.

field in the 1860's was due not to geology but to the greater scope for the introduction of improved methods afforded by a labour-force recently recruited to industrial life. But new areas formed a much higher proportion of American industry. Just as the fact that a high proportion of the demand facing American industry was new and therefore malleable demand, so the fact that American labour was less stereotyped was favourable to freedom of technical choice.[1]

But the direct influence of the character of labour-supplies on technique cannot be considered apart from their influence on the forms of industrial organisation and particularly on the relative strength of domestic industry and factory industry. It is not invariably the case that mechanisation involved the concentration of operations in the factory; in some industries highly mechanised methods could be invented in an industry organised on the domestic system and adopted in the home or small workshop, for example the sewing-machine in the manufacture of boots and shoes and clothes. But, generally, conditions in the factory were more favourable to the adoption of machines, and the general trend of technical progress was towards methods where the minimum feasible unit was too large for the domestic system. Furthermore, not merely were new methods more easily adopted in the factory than in the domestic workshop; they were more likely to be devised in the factory, since there control of production was concentrated in the hands of men who were technical experts, not merchants as in the domestic system. Thus factory industry promoted the invention as well as the adoption of new methods. Whatever circumstances strengthened the competitive position of the domestic system *vis-à-vis* the factory were, therefore, unfavourable to mechanisation.

The domestic system had certain advantages. Because of his low fixed-capital costs, the merchant capitalist was in a better position than the factory owner to cut his costs during a depres-

[1] 'The English workman, generally altogether uneducated, and with comparatively little intelligence, is, as a rule, a mere machine—a very perfect one it is true, if he be allowed much of his own way and be permitted to follow, without deviation to right or left, the groove in which he was originally started; but otherwise, he is, in too many instances, obstinacy itself, and would rather strike work than listen to anything in the form of changes and improvements.' (David Forbes, Report on Progress of the Iron and Steel Industries in Foreign Countries, *Journal of the Iron and Steel Institute* (1871), I, p. 240.

sion by reducing output, that is he could throw more of the burden of depression on to the work-people. Since he could buy and operate his equipment in smaller units than the industrial capitalist he could also increase his output more readily. Probably, too, the composition as well as the volume of his output was more flexible. On the other hand, the commercial capitalist lacked control over the quality of the product and over the performance of the domestic worker; from earliest times there were complaints of inferior workmanship and of embezzlement of raw materials. Moreover a substantial amount of circulating capital was tied up in work in progress between the commercial capitalist and the various types of domestic worker. Probably the system wasted capital as well as labour, and concentration of operations would in most cases have made better use of resources, even when it did not involve the use of a superior technique. But until the development of such techniques which could be used only in factories, the gains from concentration were not large or certain enough to elicit the entrepreneurial energy necessary to overcome the social (and in the sixteenth-century legislative) resistance to new forms of industrial organisation. Even with the advantage of superior machines the factory did not automatically supersede domestic industry. Sometimes the two systems were complementary, as when the mechanisation of iron production cheapened the raw material of the domestic workers in the Black Country nail industry. But even when the two systems competed, the domestic system had considerable powers of resistance where circumstances were favourable.

To the relative competitive strength of the two systems the terms on which labour was available to each were of considerable importance. It has been argued that in early nineteenth-century U.S.A. the conditions of labour-supply were relatively unfavourable to the growth of the factory. One of the causes of dear labour was its high rate of turnover and this was more of a disadvantage to the factory than to domestically organised industry, since the factory had heavy capital investment and needed a *stable* labour-force to make the most of it; that is an unstable labour-force hindered the factory from deriving the benefits of its labour-saving, capital-intensive equipment. This influence has been rated so highly that one American historian concludes: 'no matter what degree of standardisation the technical process of manufacturing reached, the absence of a cheap labour-supply precluded conversion to factory methods'.[1] There is obviously some truth in the argument. While labour turnover was a problem in both America and Britain in the early nineteenth century, it was more of a problem in the former and the risks of tying-up capital in bricks and mortar were, to this extent, greater. But this was only one of the ways in which labour conditions affected the balance between factory and domestic industry.

For the domestic system, quite apart from the type of techniques employed, was a labour-intensive form of industrial organisation; even in the relatively advanced form it achieved in the metal manufactures of the Black Country, a good deal of labour was employed simply in the carriage of work, as it passed through the several stages, from one out-worker to another.[2] It was only where labour was so abundant that people were prepared to work for a pittance that this system could easily survive once the factory was able to avail itself of techniques which were superior to those which could be worked in the domestic system. A general scarcity of labour therefore worked in favour of the factory, wherever it was easier for the factory to adopt compensating techniques.

This argument applies whether the workers in the domestic industry were full-time, as were most of those in the Black Country industries, or combined industry with agriculture as did many textile workers in the early days. But in the latter case there is an additional consideration. It was the rural part-time putting-out system that was most likely to impede the growth of the factory, because the labour, being part-time, was less accurately costed. The degree of resistance it put up was greatly influenced by the levels of agricultural income. In England earnings in agriculture were depressed in the 1820's and '30's, and the necessity to supplement them by part-time industry was therefore the stronger. In the U.S.A., by contrast, the creation of a market for the cash crops of New England agriculture in the first half of the nineteenth century made farming households less reluctant to give up domestic industry and buy manufactured goods from the factory; it stimulated the 'transition from mother- and daughter-power

[1] O. Handlin, *Boston's Immigrants, 1790–1865* (Cambridge, Mass. 1941), p. 81.
[2] Timmins, *op. cit.* p. 393.

to water- and steam-power'.[1]

Where labour was abundant and agricultural earnings low, workers in domestic industry reacted to the competition of superior factory methods by increasing output and lowering their rates and this retarded mechanisation. The process can most clearly be seen in the introduction of the power-loom into the cotton-textile industry. The first practical power-loom dates from the second decade of the nineteenth century, but it was not until the 1850's that the English industry went over more or less entirely to a power-loom basis. For this there are several reasons.

For the first two decades after its invention the power-loom was not unequivocally superior to the hand-loom; it saved labour per unit of output but at the cost of an increase in capital. At this stage, since additional labour could easily be obtained in England, the embryo industrial capitalists, like William Radcliffe, who in the later eighteenth century concentrated hand-looms in a single workshop, had less inducement to expand capacity by installing power-looms than would otherwise have been the case. In the second stage, by at latest the beginning of the third decade of the nineteenth century, the power-loom was clearly superior to the hand-loom; not only had improvements been made to the working of the looms, but once power had been installed in a factory the cost of operating additional power-looms was greatly reduced. Thereafter whenever a factory owner installed additional capacity he installed power-looms. But, even so, conditions of labour-supply still inhibited their installation. For alongside the factory there existed a large industry organised on the domestic system and using hand-looms. Between 1819 and 1829 the number of hand-looms fell from 240,000 to 225,000. But output per hand-loom increased, perhaps by between 25 and 30 per cent, and the prices of their products fell.[2] This was entirely the result of

[1] P. W. Bidwell, 'The Industrial Revolution in New England', *American Historical Review*, XXXVI (1921), p. 694.

[2] The estimate of the increase in output per hand-loom is based on the following figures.

	Hand-looms	Power-looms	Index of output	Index of capacity (a)	Index of capacity (b)
1820 .	240,000	14,150	100	100	100
1829 .	225,000	55,500	180	138	145

Capacity (*a*) is based on the assumption that one power-loom was equivalent to

the application of more labour to each hand-loom, not of any technical improvements. 'The hand-loom weavers', wrote the Commissioners, 'are forced to offer their services at a rate of wages which render them rather cheaper than those of the power-loom: and which can continue only until some further improvement shall again have made the power-loom a successful rival, and the hand-loom can be kept at work only by a still further reduction; and thus the unequal race continues...'.[1] The result of this was that a large amount of labour continued to be tied up in hand-loom weaving.

But this response of the domestic system also hampered the factory sector of the industry. In the first place, the increased production of the hand-looms caused prices of cloth to fall more rapidly and the profits of the factory sector to be lower than would otherwise have been the case, thus reducing the ability and the incentive of factory owners to install more looms. In the second place it offset in some degree the technical superiority of the power-loom to the hand-loom. Improvements in the productivity of the power-loom as a result of labour- and capital-saving inventions were to some extent matched by improvements in the productivity of the hand-loom as a result of applying more labour at lower earnings. Because hand-loom weavers were prepared to tighten their belts, the labour costs of hand-loom production fell more than those of power-loom production.

The fall in the price of hand-loom production probably did not exert much influence on the choice of technique within the factory, at least after Roberts' improvements; though it temporarily reduced the degree of superiority of the power-

three hand-looms in both 1820 and 1829. This is the equivalence given by Babbage (quoted by Porter, *Progress of the Nation* [1836 ed.], I, p. 233) and in the S.C. on Manufacturing Employment, P.P. 1830, x, p. 4. Capacity (*b*) is based on the assumption that one power-loom was equivalent to two hand-looms in 1820 and to three hand-looms in 1829 (Ellison, p. 69, gives the rate of production of a power-loom as 600 lb. in 1820 and 900 lb. in 1830). R. Guest in *A Compendious History of the Cotton Manufacture* (Manchester, 1823), pp. 47–9 says that 200 steam-looms produced as much as 875 hand-looms.

Hand-looms from Ellison, *Cotton Trade*, p. 66; power-looms from Chapman, *op. cit.* p. 28, fn. 1. There is a margin of error in both estimates. Output from Ellison's figures of raw cotton consumption (Statistical Table No. 1).

On assumption (*a*) the output per hand-loom weaver increased between 1820–9 by approximately 30 per cent, and on assumption (*b*) by approximately 25 per cent.

[1] R.C. on the Hand-loom Weavers, P.P. 1841, x, p. 25.

loom it by no means eliminated it. But among factory owners there were some who employed domestic hand-loom weavers, in addition to workers on power-looms within the factory; and the increase in output at lower prices of the domestic producers delayed an increase in power-looms by such employers. The reduction in the price of hand-loom production also reduced the urgency of improving existing power-looms. Moreover it exerted an influence on the wealthy commercial capitalists who provided work for the domestic weavers. Few of these became industrial capitalists. Perhaps they lacked the technical expertness to do so; perhaps they already had such a wide range of functions that to add another would have been difficult. Yet if the hand-loom weavers had not been so willing to produce more at lower prices, these commercial capitalists would have had a much greater incentive to withdraw their capital from the domestic system and devote it to the building and equipment of factories.

Thus the existence of a large sector of the weaving industry organised on the domestic system acted as a brake which periodically pulled up the mechanisation of the industry. Under the stimulus of increased demand or because innovation reduced the cost of power-loom products, there was investment in capital-intensive plant. But, after the boom caused by this investment, there would be a fall in prices, and, if there had been cost-reducing innovations, a fall to a lower level than before. This reduced the income of hand-loom workers, who aggravated the situation by increasing output. This shifted the profitability of investment towards a labour-intensive technique and there might be investment in the hand-loom industry, perhaps by non-innovating capitalists.

While the existence of an established industry organised on the domestic system provided recruits for the factories, it also constituted an impediment to the mechanisation of the cotton-textile industry in England which was not present in an area where the industry was organised from the start on a factory basis. A domestic system could, of course, exist without cheap and abundant labour; it existed for example in the Rhode Island branch of the American industry and around Philadelphia. The hand-loom weavers were a problem in New England as well as in old.[1] But their numbers were smaller

[1] N. J. Ware, *The Industrial Worker, 1840–60* (Boston, 1924), pp. 61–2.

and their resistance much less. This was partly because of the difference in the timing of development in the two countries to which we have already referred; because in England it was technically possible to mechanise spinning at an earlier stage than weaving, the mechanisation of spinning led to a large increase in domestic weaving. But if labour had been scarce in England, the domestic system would have contracted much more rapidly than it did. It was because it was difficult to get other jobs that the natural increase among the families of the hand-loom weavers doubled up on the hand-looms and endured so severe a reduction in their earnings. It was in domestic industry that the overspill of surplus agricultural labour tended to be most severely felt. 'This excess of supply' said a witness before a Parliamentary Committee in 1854 'arises primarily from the accessibility of the trade of framework knitting to the unemployed labour of all other classes, and from the facilities with which a knowledge of the trade, especially in the common branches is acquired. . . . These facilities admit of the competition of women and children, of which they avail themselves to a very considerable extent, the natural consequence of which is to reduce wages generally, by an undue augmentation of numbers in the labour-market, and particularly those of male adult knitters, by lowering them to the standard rate of such competitors'.[1]

Where the domestic industry was urban and its workers skilled and full-time, the relations with the factory were in many respects different, and in certain cases—in boots and shoes and in paper for example—there developed a specialisation of function between the two types of industry. This was possible where trade unions maintained the skill of the handicraft labour, where the machines were not capable of making products of the highest quality and therefore allowed a distinct differentiation between hand-made and machine-made goods,

[1] S.C. on Stoppage of Wages (Hosiery), P.P. 1854–5, xiv, Qu. 776–8, 2573. The system of poor relief may also have helped to sustain the domestic system against the factory. In the industrial areas the people who relied most on outdoor relief were the domestic workers; relief within the poor house was not suitable for their condition in times of distress, for they were impoverished not by unemployment but by low prices, which might indeed cause them to work harder. It was better to give them aid by, for example, paying their rents rather than force them to sell the tools on which their livelihood depended or allow landlords to distrain on work in progress. There are signs that even after the new poor law of 1834 poor relief operated as a subsidy to domestic workers.

and where there was a market prepared to pay for quality products. In the case of such industries, the Webbs argued, the maintainance of high rates of pay for handicraft labour gave 'the utmost possible encouragement to the introduction of machinery, wherever machinery could possibly be employed'. But the conditions for such a division of labour were not widespread, and in the chain and nail trade of the Black Country and among tailors and cabinet-makers, the 'constant yielding of the handworker . . . diminishes the pressure on his employer to adopt the newest improvements, and positively tempts him to linger on with the old processes'.[1] Perhaps the most striking example of the influence of abundant labour in facilitating the survival of a domestic industry comes from one which was urban and involved a high degree of skill, the manufacture of small arms. At the middle of the century the American small-arms industry was already a highly-mechanised industry using interchangeable parts. But in Birmingham, the greater part of the gun, with the exception of the barrel of the musket, was produced by hand-labour alone; the different parts were manufactured separately and then a viewer saw to it that they were set up accurately. The cost of a musket produced in Birmingham was estimated by the government as a little under 60 shillings as against 37 shillings for a comparable American product. Hardly any machinery was employed in the production of bayonets in England and consequently, reported two government inspectors, 'the expense of a bayonet was considerably more than double what it might be expected to be, made by machinery'. In this case there was no one within the Birmingham industry who thought it worth while to devise machinery or to adopt the machinery available in America: the machinery was ultimately introduced by the State. 'It was a question', said Gladstone, defending the proposal to do this, 'between superior and inferior processes, between labour and machinery'.[2]

SKILLED VERSUS UNSKILLED LABOUR

But granted that labour in general was more abundant in England than the U.S.A. in the early nineteenth century, skilled labour in relation to unskilled was scarcer in England.

[1] B. and S. Webb, *Industrial Democracy*, pp. 422, 416.
[2] *Hansard*, February 1854, pp. 1402–3.

Skilled labour was, of course, dearer than unskilled, in both countries; and in both countries this provided manufacturers with an incentive to adopt and devise methods which replaced skill by non-skill. But in England elasticity of supply to the individual firm of unskilled labour, in relation to skilled, was greater than in America before the start of heavy immigration. Where types of labour are not easily substitutable for each other and differ a great deal in their responsiveness to demand, there is an incentive to invent and adopt machines which economise in the unresponsive type. So that while the scarcity of labour in general relative to capital provided the Americans with a stronger incentive than the English had to replace labour by capital, the English had a stronger incentive than the Americans to replace skilled by unskilled labour. We have considered the effect on techniques of the composition of the labour-force in America, particularly after the beginning of heavy immigration in the 1840's and we shall discuss the same problem in the English context.[1]

The strength and fruitfulness of this incentive depended on (*a*) the possibilities of alleviating a shortage of skilled labour by other methods than mechanisation and (*b*) the technical possibilities of alleviating it by mechanisation.

In the long run a shortage of skill could be alleviated, at a given level of mechanisation, by the substitution of adjacent grades of skill, by training, and by increased division of labour, that is by breaking down the operation requiring skill into component parts capable of being performed by unskilled labour. While, as Adam Smith argued, division of labour may facilitate the invention of new mechanical methods, it was also an alternative to mechanisation, for example in the manufacture of metal goods in the Birmingham area.

But there were certainly shortages of some types of labour which persisted at least over the course of a single trade-cycle, and where the skills were not only highly specialised but were protected by craft regulations, the supply of them might be completely inelastic for several years. In these circumstances no significant expansion might be possible without mechanical means to replace skilled by unskilled labour.

There are several instances where the desire to diminish the bargaining power of skilled craft labour provided a strong

[1] See the discussion on pp. 129–131 above.

incentive to install machines which replaced it by unskilled labour. The adoption of the Fourdrinier machine after 1800 by English paper manufacturers was due largely to their desire to break the power of skilled labour in the industry.[1] The invention of the automatic spinning-mule by Richard Roberts in 1825 was prompted by the request of a committee of cotton manufacturers of Manchester who found the mule-spinners intractable, and Glasgow cotton manufacturers in 1833 ascribed the introduction of the self-actor mules into their mills to the desire to avoid the use of skilled labour.[2] The stimulus in this case was the strong bargaining position of the highly-skilled adult spinners, who worked the old hand-mules, compared with the less skilled labour which was competent to work the self-acting mules. The experiments of the Birmingham manufacturer, Osborne, which led to the invention of a means of making gun-barrels by grooved rolls, were occasioned by a strike of barrel welders.[3] According to Samuel Smiles 'in the case of many of our most potent self-acting tools and machines, manufacturers could not be induced to adopt them until compelled to do so by strikes'; and besides the self-acting mule he instanced the wool-combing machine, the planing machine, the slotting machine and Nasmyth's steam arm.[4]

The real question is how common were the cases where the advantage of a method depended on the extent to which it replaced skilled by unskilled labour, as distinct from replacing labour in general by capital? According to Andrew Ure, the new techniques of the early nineteenth century characteristically replaced skilled by unskilled labour and this was the main motive for their adoption in England. 'On the handicraft plan', he wrote, 'labour more or less skilled, was usually the

[1] D. C. Coleman, *The British Paper Industry, 1495–1860* (Oxford, 1958), ch. VII, pp. 258–9.

[2] Roe, *op. cit.* p. 51; Chapman, *op. cit.* p. 69; P.P. 1833, VI, Qu. 5397.

[3] Timmins, *op. cit.* p. 389. As a result of a strike of the boiler-makers in his Manchester works in the later 1830's, William Fairbairn invented a successful riveting machine, and in 1873, in old age, he generalised his experience, and wrote that 'the introduction of new machinery and the self-acting principle owed much of their efficacy and ingenuity to the system of strikes, which compelled the employers of labour to fall back upon their own resorces, and to execute, by machinery and new inventions, work which was formerly done by hand'. *The Life of Sir William Fairbairn*, edited and completed by William Pole (1877), pp. 163–419.

[4] S. Smiles, *Industrial Biography* (1863), pp. 294–5.

most expensive element of production but on the automatic plan, skilled labour gets progressively superseded ...'. And again, 'it is, in fact, the constant aim and tendency of every improvement in machinery to supersede human labour altogether, or to diminish its costs, by substituting the industry of women and children for that of men; of that of ordinary labourers, for trained artisans'.[1] The outstanding instance of such an improvement was the water-frame which was essentially a substitute for human skill.

The improved machines, however, almost invariably required more skill to make them than did the simpler machines; and where it was impossible to substitute unskilled labour in their production the greater inelasticity of supply of skilled labour raised the price of machines (relatively to general labour) including the price of those machines which in their operation substituted unskilled for skilled labour. In the operation of the new machines too there were some which did not conform to Ure's description and which required more rather than less skilled labour per unit of output. 'Jennies and mules could be worked economically only by skilled spinners, and their effect was therefore simply to reduce the quantity of labour needed for a given output and to substitute men's labour for women's and children's, because greater strength was required to use them than the one-spindle wheel'. Partly because hand-loom weaving was an occupation for men who did not take easily to the factory, the first weavers on the power-looms were nearly all women; but when the power-loom was more widely adopted much of the work on them was done by men, and there was probably not much difference of skill between hand-loom and power-loom weavers.[2] In the case of the iron and steel industry it is more difficult to assess the quality of labour required by different types of operation but in the 1870's the lack of 'a highly intelligent class of workmen to carry out the practical details' was held responsible for the fact that many valuable inventions in the steel industry had been abandoned in England;[3] presumably the new methods required more skilled labour per unit of output than the old, though it may simply be

[1] Ure, *Cotton Manufactures*, II, pp. 155–6.

[2] Chapman, *op. cit.* pp. 69, 54.

[3] John Jones, 'Technical Education in Connection with the Iron Trade', *Journal of the Iron and Steel Institute*, 1876, pp. 343–4.

that the minimum size of operation required a very large amount. There were, therefore, certainly a number of cases in which the more mechanised methods did not alleviate a shortage of skill.

Finally a scarcity of skill induced by trade union regulations was not on a par with a natural scarcity of skill. Quite apart from the fact that the former entailed costs for the economy as a whole, the manufacturers reacted in a different way. Nasmyth complained, from his own experience, that before an employer 'can introduce even some of the most obvious improvements he will have to consider "Will my men like this; will they turn out?" '.[1] Moreover where there grew up detailed agreements about piece rates any change of technique involved a recalculation of the rates and therefore threatened to impair relations between employers and workers. A stoppage was worth risking only where gains from the new technique were certain to be considerable and in the interests of good labour relations changes which would otherwise have been advantageous were not made. The existence of a standard list by which the price of labour was regulated in the English cotton-textile industry made it extremely difficult to make variations in the piece rate according to the number of machines run by the operative. Young tells of a Lancashire mill at the end of the century which tried the experiment of giving two warpers three mills between them instead of one each, but abandoned it because, although the warpers earned better wages, they did not get the full production out of the machines. 'When I explained this to an American manager, he could not at first believe that the Englishman was not making a foolish mistake. "Why," he said, "don't they buy another frame, or two more, or as many as may be necessary to make up for the diminished output? The cost of a frame is only so much"—he named a figure—"and they would save as much in labour"—here he made a rapid calculation—"as would pay for the additional machinery in a very short time. And the warpers might still receive much higher wages." ' 'I reminded him', wrote Young, 'that the price of labour in Lancashire cotton mills was regulated by a standard list which did not provide for variations in the piece rate according to the number of machines run by the

[1] R.C. on Trade Unions, 1867, Fifth Report, P.P. 1867–8, XXXIX, Qu. 19222.

operative. 'Ah,' said the American, 'that accounts for it'.[1]

There are other cases where the attitudes of the unions retarded mechanisation. The first mechanical puddling machine was devised as early as 1836, but not until the later '60's were large-scale attempts made to replace puddlers. These attempts were undoubtedly stimulated by the unionisation of the iron workers in the '60's which roughly coincided with heavy investment in the industry. But the unionisation which stimulated the attempts to mechanise also contributed to frustrate them: so long as investment in furnace capacity was increasing, puddlers were so much in demand that their bargaining position was strong. 'So long as there are more furnaces than men, the puddler will simply leave the master and his improved furnaces and work at the place where he finds all things in the old style in which he has been educated and where he is least interfered with'.[2] Resistance by unionised employees in the Staffordshire iron industry in the 1860's compelled the employers to pay the same piece rates on new machinery to the 'shinglers' or 'nobblers' as they had on the old hammers, despite the fact that the new machinery was very labour-saving.[3] In general, the unionisation of skilled workers in the 1850's and '60's, while it accentuated the unresponsiveness of the supply of skilled, compared with unskilled labour, also made it more difficult to thwart it by mechanisation.

The fact that it was more difficult in England to reduce piece rates to the extent warranted by the increase in productivity of labour due to labour-saving machines made such machines less attractive to the English manufacturer. The inflexibility of piece rates also impaired their usefulness as an instrument for capital-saving; in the U.S.A. a reduction of piece rates was one way in which manufacturers attempted to speed up their machinery.[4]

On the whole therefore it seems doubtful whether the need in England to replace skilled by unskilled labour was a stimulus to mechanisation at all comparable with the need to replace labour by capital in the U.S.A.

[1] T. M. Young, *op. cit.* p. 29.

[2] *Journal of the Iron and Steel Institute*, 1882 I, p. 168.

[3] P.P. 1867–8, XXXIX. Qu. II, 141–11, 158. For opposition to the mechanical furnace see *Journal of the Iron and Steel Institute*, 1874, pp. 494–5.

[4] N. J. Ware, *op. cit.* p. 111.

NATURAL RESOURCES

More difficult problems arise when we consider natural resources. Granted that labour was not a restraint and a stimulus in England to the same extent as it was in the U.S.A., the supply of land and possibly of certain other natural resources, imposed a greater restraint. Why should such shortages be less effective in stimulating technical progress than a shortage of labour?

The desire to economise scarce natural resources—to alleviate shortages that arose during the process of capital accumulation—was obviously a powerful source of technical progress. The high price and inelasticity of the supply of land in England did lead to a wider adoption in the late eighteenth century of those known techniques—the new rotations with root crops—which saved land (that is increased output per acre), and the adoption of these methods did put English farmers in a position to make improvements which more than compensated for the cost and inelasticity of English land (as well possibly as yielding some increase in output per head.) The factor-endowment of the two economies biased their agricultural investment in different directions. As Jefferson wrote, 'In Europe, the object is to make the most of their land, labour being abundant: here it is to make the most of our labour, land being abundant'.

Fuel and power are perhaps the most homogeneous natural resources and thus the two most comparable to labour. Shortage of both was an incentive to invention in England. Many of the mechanical improvements to mills and millwheels in the early English industrial revolution resulted from attempts 'to make the most of a given stream'.[1] Much of the history of innovation in the English primary iron industry in the eighteenth century can be explained by reference to fuel on the same lines as American developments in the nineteenth century have been explained by reference to labour. Fuel was dearer and less elastic than in other countries; the imperfection of the product market ensured high average profits and a low marginal rate on capital. Imports from Sweden and Russia set a limit to the extent to which rising costs could be passed on in high prices. Periodic wars produced a sharp increase in demand

[1] John Banks, *Treatise on Mills* (1795).

for iron and gave periods of high, though temporary, protection. Ironmasters, therefore, had an incentive to search for fuel-economising methods and found them by replacing dear and inelastic charcoal by coke, which was cheaper and the supply of which was more elastic. This explains why the new methods were discovered in England rather than in Russia, and also why the American iron industry was slow to turn to coke.

The new methods of producing iron were expensive in fuel, and because they were new and imperfect some of the expense was unnecessary; they were wasteful of coal and visibly so since the gases could be seen to be escaping during the process. English ironmasters, therefore, had a strong incentive to explore the possibilities of economising in fuel. Throughout the nineteenth century, from the Hot Blast to the Bessemer process fuel economy continued to be a most powerful motive behind technological innovation. Fuel economy was the purpose of Watts' first improvement to Newcomen's engine (the patent of 1769), and continued to be the motive behind developments in the steam engine from Watts' separate condenser down to the compound engine. In its early days the characteristic of the steam engine was that it increased output per unit of power but at some increase in capital per unit of output. It was therefore first adopted in areas and industries where power was dear and its supply inelastic, for example where supplies of water-power were small and irregular. It was the small uncertain supply of water-power at Birmingham which made Boulton give so much encouragement to Watt in the first place.[1] The inelasticity of raw materials and foodstuffs directly stimulated innovation in transport. The desire to save fuel was the motive behind the development of the regenerative principle by C. W. Siemens.[2]

Many of these resource-saving innovations initially did no more than compensate for the dearness and inelasticity of natural resources in England, but ultimately they were all developed to a point where they increased outputs for all

[1] H. Hamilton, *British Brass and Copper Industries* (1926), p. 125.

[2] The development of the regenerative principle by C. W. Siemens was inspired by desire to use fuel more economically, and it was applied to puddling because this process, as then practised, was extremely wasteful in iron and fuel. W. Pole, *The Life of Sir William Siemens* (1888), p. 50.

factors. Thus a moderate range of innovations of English origin can be attributed to natural-resource scarcity in England, on the same line of argument as the characteristic American inventions can be explained by reference to labour-scarcity. We might expect, from this analysis, simply some difference in type of technology, but no difference in the rate of technical advance.

There is no way of balancing the effect of those British inventions which sprang from the limitations of natural resources against the American inventions that sprang from labour-scarcity. There is, in principle, no reason for supposing that it is easier to relax a shortage of natural resources without recourse to resource-saving techniques than a shortage of labour without recourse to labour-saving devices, no reason in principle why we should expect it to be easier to replace a man than a natural resource by a machine. There is, however, a reason for expecting that attempts to save natural resources should be less fruitful—in the sense of producing a smaller increase in output over the whole economy—than attempts to save labour. For since labour is more homogeneous than natural resources, a shortage of labour provides a more general incentive—an incentive to more industries—to invent techniques to offset it, because all firms use a great deal of labour, but only some use much of any particular natural resource except power.[1] Since the range of types of labour is more continuous than that of kinds of raw material, inventions which save labour are likely to be more widely applicable, or to suggest possibilities of new methods over a wider range of processes than are those which save specific raw materials. The method of interchangeable parts is a case in point. The contrast is not an absolute one—a fuel shortage can be general and a fuel-saving invention (for example the steam engine) can benefit many industries—but labour is common to more industries even than fuel.

There is also perhaps another reason of a general kind. Natural resources were usually a smaller proportion than labour of total costs, so that the same percentage rise in costs was more important when it took place in labour. There may be some

[1] It might be argued that shortages are more likely to stimulate invention when they are specific than when they are general. But there are not many cases of proved relationship between invention and raw material shortage. There is one from cotton; the cotton famine during the American civil war stimulated attempts to improve the machinery for spinning short-staple cotton, and accelerated the adoption of the self-actor mule (Chapman, *op. cit.* p. 70).

illusion here: the fact that labour was a high proportion of the cost may have induced industrialists to draw the deduction, which does not necessarily follow, that it was the cost that was most easily reduced.

There are, therefore, general grounds for supposing that there was an asymmetry in principle between the effects of natural-resource scarcity and labour-scarcity. But, in any case, there are reasons for supposing that there was one in practice in the conditions of the first half of the nineteenth century. A rise in the price of natural resources stimulated a search for new sources of supply which tended to countervail the incentive to invent, and this search was probably more active than the search for additional supplies of labour induced by labour-scarcity, and more successful. Men of enterprise have an interest, when for example minerals are scarce, in prospecting for new beds of ore and in developing them. The supply of free labour is a less appropriate or profitable task for private enterprise. Scarce materials can be imported from wherever they are in good supply much more readily than labour, at any rate since the abolition of the slave trade. Thus in the first half of the nineteenth century it was probably easier to relax natural-resource stringency in England, simply by tapping new sources of supply and without technical changes in the industries using the scarce resource, than it was to relax labour stringency in the U.S.A.[1]

It is also probable that the technical knowledge of the nineteenth century was more capable of providing solutions to problems of labour-scarcity than to those of natural-resource scarcity. It may be that this was partly because much early nineteenth-century technology had been developed to meet problems of labour-scarcity and so was attuned to this particular demand. But the main reason is that most solutions of natural-resource scarcity problems have been obtained from chemistry and electrical engineering, and in the nineteenth century these lagged behind mechanical engineering, simply because they are more difficult subjects. It is possible, too, that a labour-saving bias to invention was ultimately favourable to the adoption of resource-saving inventions made elsewhere;

[1] Though, of course, technical change might be necessary to tap a new source of supply; the expansion of cultivation even in the U.S.A. necessitated new agricultural techniques, especially on the Great Plains.

for it was favourable to the development of mechanical skills which were a prerequisite for the adoption of new methods whatever factor they saved.

CAPITAL

We shall now consider the effects on technical progress of the relative shortage of capital in England. If labour was abundant in relation to capital in the U.K., did not this encourage British industrialists to devise methods of replacing capital by labour just as much as the relative scarcity of labour encouraged Americans to devise methods of replacing labour by capital? Did not the English industrialist have an incentive to make the best use of his scarce machines as great as the American's incentive to make the best use of his scarce labour? Why should the desire to save capital not be as fruitful a stimulus to technical development as the desire to save labour?

It is evident that a good deal of inventive ability in England was in fact devoted to improving the quality of the capital equipment. Indeed, in his discussion of inventions, Adam Smith devoted most attention to those which saved capital rather than labour; he wrote of 'improvements in mechanics which enable the same number of workmen to perform an equal quantity of work with cheaper and simpler machinery than had been usual before'.[1] And, as Ashton has shown, several of the improvements in the eighteenth century were of this nature, particularly those which economised horse-power.[2] The attraction of coke in place of charcoal in the primary iron industry was partly that it enabled the furnaces to be run for longer periods—the charcoal blast lasted only about twenty-three weeks in the year. One of the attractions of replacing water by steam was the greater regularity with which the steam-driven machinery could be used. In the agitation against the Factory Acts, the objection of many English manufacturers to the shortening of hours by legislation was not so much that it would increase labour-costs per hour—because many manufacturers expected wages to fall in much the same proportion as hours—but that it would limit the hours they could run their machinery and so increase capital-costs per unit of output. What Sir

[1] Adam Smith, *Wealth of Nations*, Book II, ch. II, p. 271.

[2] T. S. Ashton, *The Eighteenth Century*, p. 110.

James Graham objected to was that legislation would 'abridge the hours your machinery is to run by no less than one-sixth part'.[1] Though the use of iron and steam in ships obviously reduced the labour employed per unit of cargo carried, its principal advantage was that it made more effective use of shipping capacity, by enabling larger, faster and more durable ships to be built and reduced their costs of maintenance. It has also been suggested that when account is taken of the economies they effected in working capital, the railways were capital-saving.[2]

Thus there were improvements which were principally capital-saving. But there are reasons why a desire to save capital should have been a lesser stimulus to improvement than desire to save labour.

The principal point here—indeed the principal point in the whole argument—is that relative abundance of labour means the absence of a restraint on the rate of profit, and, in so far as capital accumulation depends on the rate of profit, on capital accumulation; the relative scarcity of labour means the converse. Any industrialist has an incentive to devise and adopt methods which reduce costs, but the need to avoid a falling rate of profit is a more powerful incentive to devise new methods than the possibility of increasing the rate of profit. Moreover, while inelastic supply to the firm of *any* factor of production is likely to stimulate innovation, a capital-shortage is less likely to be a continuous stimulus to innovation than a labour-shortage. A successful innovation, of whatever type, by definition increases profits, and these provide more internal finance and also improve the firms' ability to borrow. A successful innovation *may* also improve the ability of a firm to command labour, but the supply of finance is more directly responsive to successful innovation than the supply of labour. So that random improvements of a neutral character will tend to alleviate capital-shortage more than labour-shortage. The point here is that Americans had a stronger incentive to try to find new methods—in *any* direction. The *particular* bias which the relative scarcities of capital and labour gave to technical progress in the two countries was also more favourable in the U.S.A. than in Britain.

[1] *Hansard*, 3rd series, LXXXV, pp. 1245–6.

[2] J. Robinson, *The Rate of Interest and other Essays* (1952), p. 43.

There are several reasons for supposing that on balance the search for labour-saving devices is likely to be more fruitful than the search for capital-saving devices. The principal reason is that the desire to save labour drives the manufacturer of a particular product towards the more capital-intensive of existing methods of production and these are the more complicated, the more highly mechanised techniques. It is necessary, at this point, to guard against a possible misunderstanding. The argument is not that, over the whole of industry, the most capital-intensive processes are the most highly mechanised: this would be obvious nonsense since many of the most advanced (in the sense of highly mechanised) techniques use very little capital in relation to output. We are concerned only with alternative techniques for producing the same commodity, and only with these when the choice between them is influenced by the relative price of factors. The argument is that, within *this* range of techniques, the more capital-intensive is the more highly mechanised; it therefore affords better opportunities for the acquisition of know-how and greater intrinsic technical possibilities of further development. Within this range the capital-saving are the simple, the labour-saving the complicated machines: technical progress moves from the labour-intensive to the capital-intensive end of the spectrum. This is so simply because, in the advanced countries, capital has accumulated more rapidly than population. It is therefore the earlier techniques of an industry that are the most labour-intensive; that is, the possibilities of the labour-intensive end of the spectrum are the ones which are explored first in time, and it is the capital-intensive end which points in the unknown direction.

The new and more complicated techniques which resulted from the search at the end of the capital-intensive end of the spectrum, almost invariably, when they were fully developed at least, saved capital as well as labour and many saved more capital than labour: the effects of new techniques on factor-proportions often had little relation to the motives which prompted the search, since invention is a chancy business. This is to say that there have been a large number of inventions the *incidental* effects of which were capital-saving. But the effects of a decision to adopt the more capital-saving of existing techniques and to search consciously for new capital-saving methods were different. Such capital-saving decisions were likely to result in the adoption of simpler techniques, machines from which there was nothing for others to learn; in the extreme case, for example, to some form of shift-working from which nothing new was to be learned, except possibly some lessons in labour management. This is fundamentally the reason for supposing that the techniques which replaced labour by machinery lent themselves to improvement more easily than those which replaced machinery by labour.

Moreover, capital- and labour-shortage show themselves to the individual entrepreneur in different ways, and he does not react to them in the same fashion. A shortage of labour reveals itself to an established firm in a rise in the wages and a threatened diminution in the size of its labour-force. If the entrepreneur does not take action to increase the productivity of his labour, he will have to contract his operations. But a shortage of capital does not show itself in a *diminution* of the capital of an existing firm; if additional capital is very expensive or unobtainable, all this means is that the firm does not expand. Shortage of capital would more closely resemble shortage of labour if shareholders were able to demand from a firm the issue value of their shares. But shortage of capital does not show itself in the form of a withdrawal from existing firms. So far as *incentive* is concerned, therefore, a shortage of labour provides a more compelling reason to search for new methods than does a shortage of capital. But there is the additional point that it is easier to increase the productivity of an existing labour-force than of an existing stock of capital goods, for the simple reason that labour can much more easily be re-allocated. In retrospect the historian can see many inventions which were predominantly capital-saving in character, but a great many of these were external to industry, for example improvements in transport, and many others rose from improving the performance and lowering the cost of machines which, when first devised, were essentially *labour*-saving. Except in the capital-goods industries, the scope for capital-saving technical improvement in established firms was very limited. The most direct way would have been for the firm to sell the most capital-intensive section of its existing equipment, but this was rarely feasible. Something could be done by using factory space more economically; machines might sometimes have been doubled up, but only in times of boom when additional factory space could not be acquired quickly enough; and the usual motive for changing the ratio of machines to space was to save labour not capital. An existing firm which is short of capital is likely to attempt to relieve the shortage by ways which do not involve technical invention, for example by running down its stocks, or by amalgamation with other firms. An entrepreneur starting a new concern, is of course in a different position. He will try to make his capital go as far as possible and will be free to employ the most capital-saving techniques. Even in this case, however, shortage of capital is likely to affect less the techniques used than the choice of industry and the method of organising operations; a new entrepreneur who is short of capital will enter the least capital-intensive section of an industry and employ methods of organisation which make his capital go furthest, for example leasing (as opposed to buying) the most capital-intensive equipment and/or farming out the most capital-intensive operations.

Furthermore, even for the economy as a whole, a shortage of finance is more easily alleviated without technical improvement than a shortage of labour. Changes in the organisation of credit, changes in reserves policy, the development of industrial banking can alleviate capital shortage to a degree which changes in organisation can less easily do for labour. If for a moment one rules out the possibility of augmenting the resources of one economy by calling upon those of others better supplied, it is easier to augment finance than labour without recourse to mechanical improvements. An economy may have to pay for alleviating a shortage of finance by credit creation, since this accentuates other sources of instability and raises prices, though to a marked degree only where labour is scarce, and where land and labour are abundant it may have a stimulating effect. Moreover funds are geographically more mobile than labour. A remote continent of abundant land in the early nineteenth century found it easier to draw on the surplus capital of the settled regions than on their surplus labour.

A shortage of finance for an economy as a whole is an incentive to innovation, but to innovation in the capital-market rather than to mechanical innovation. Innovation in the capital-market may be as important as mechanical innovation in increasing income per head, and there is no reason, in principle, to suppose it is any easier. It took much longer to produce a stable banking system in America than in England and it can be argued that the English political system was as encouraging to innovations in finance as the American social system was to mechanical invention. The principal difference between financial innovation and mechanical innovation seems to be that the former is more likely to give rise to external economies for other countries. Although England benefited from the external economies resulting from American mechanical innovations, for example the cotton-gin in the early nineteenth century, cheaper tramways in the later nineteenth century and cheaper electric motive power in the early twentieth, these benefits were less significant than those which America in turn obtained from England's stable financial facilities.

The asymmetry between labour and capital can be expressed in another way. Idle capital is less likely to remain unemployed than idle labour. For the capital-market is less imperfect than the labour-market: capital is more homogeneous and does not have a floor set to its price such as the cost of subsistence and trade union action set to wages. If liquidity preference does set a floor to the rate of interest, the idle funds are simply eliminated by a fall in income. But before this point is reached the holders of idle funds are likely to search hard for opportunities of employing them. This means that technology is more likely to be sensitive to factor-endowment where labour is scarce in relation to capital, than where capital is scarce in relation to labour.

From the point of view of industrial relations the attempt to save labour is likely in some circumstances to meet less resistance than the attempt to save capital. In an industry the demand for whose product is inelastic and whose labour is specific, the equilibrium position in a capital-saving method probably involves some lowering of the wage rate; that is the entrepreneur will employ more workers per machine at a somewhat lower rate. The equilibrium position in a labour-saving method involves the employment of fewer labourers, but at some increase in the wage rate. Both shifts in technique are likely to cause difficulties in industrial relations, but the first may cause more difficulties than the second.

Finally, as we have already suggested, it is probable that the adoption of known techniques which saved labour with some increase in capital placed manufacturers in a more favourable position for making certain types even of capital-saving improvement than did the adoption of techniques which saved capital with some increase in labour. Where there was a large amount of machinery per worker, the manufacturer had both a greater incentive and more opportunities for improving the performance of machines. The scope for capital-saving inventions was particularly great where the labour-saving innovation involved big changes in machinery—the introduction of much more complicated machines. There were then a large number of technical possibilities for improving the machinery—reducing its cost per unit of output. The design of the new type of machine could be tidied up, and what turned out to be unnecessarily complicated ways of doing things simplified; and engineers could search for ways of increasing the efficiency of operations. But quite apart from the possibilities of such improvement, there was likely to be a fall in the price of labour-saving machines once they had been widely adopted; a new type of machine on its first introduction in a limited number of firms was often very expensive and its price came down only when it was produced in large numbers and enjoyed the economies of scale. Thus capital-saving often naturally followed a labour-saving innovation, especially if the latter was made in a boom, since the fall in demand during a slump stimulated attempts to cut costs, and the lower wages and greater availability of labour made small improvements and simplification of machines and running them longer the obvious way to do this.

It is also true, of course, that important capital-saving improvements were made in developing inventions which in origin were principally natural-resource saving. The most conspicuous examples are the improvements in the effectiveness of the steam engine. The development of the high-pressure engine by Trevithick made it possible to build engines which were much cheaper per unit of power. In the 1840's pressures were increased by the addition of new boilers to the same engine; engines of the Watt type were converted to compounding by adding high-pressure cylinders at the end of the beam. Moreover, improvements in the application of power to transport saved space and reduced capital-costs by reducing the volume of goods 'in the pipe-line'. But if, as we have argued in an earlier section, labour-saving inventions were commoner than those which saved natural resources, the Americans were to this extent in a better position to make consequential capital-saving inventions.

The Americans also had some advantage in the search for methods of reducing the cost of making machines of a given performance. From the point of view of the economy as a whole, probably the most important capital-saving inventions were labour-saving inventions in the making of machines.[1] The U.S.A. was in a better position than the U.K. to make these labour-saving developments in the making of machines for exactly the same reasons as it was in a better position to make them in the using of machines. In the early stages the makers of machine tools had their problems set for them by the machines invented in other fields. Thus the great advances in English machine tools from 1775 to 1850 came from attempts to improve the manufacture of mining-equipment, steam engines, rails and ships. The Americans took over most of these advances. The problems they had to face for themselves were somewhat different. They had to develop tools which would replace labour. The methods specifically designed to replace labour in one operation, for example the system of interchangeable parts, were widely applicable to other operations and, to some extent, they were applicable to the manufacture of machines themselves.

There is finally a general reason why a technology with a present bias towards labour-saving should yield greater future gains than one with a bias in some other direction, a reason which is quite distinct from the fact that, at any given time, it is the capital-intensive end of the spectrum which has the greatest technical possibilities of improvement. As economic growth has in fact proceeded in the west in the last hundred years or so, the terms of trade between labour and capital have moved in favour of the former. Except possibly in England in

[1] 'The high price of labour makes labour-saving the first and main consideration (that is in the design of woodworking machines); cheap construction, bad taste in design, rough fitting, are all lost sight of; it is the quantity of stuff that can be "got through" in a given time that decides the merit of a machine.' (Richard's *Treatise*, pp. 52–3.)

the early decades of the nineteenth century, population has grown at a slower rate than capital, and the labour-force has grown even more slowly, even allowing for increases in the quality of work with the fall in hours worked. The return to capital has therefore fallen relatively to that of labour; the return to capital is much the same as it was in the eighteenth century and wages are much increased. There has been a fairly stable capital–output ratio, but a constant rise in the proportion of capital to labour.

It is clear also that natural resources as a whole have increased more rapidly than labour. The movement in the terms of trade between capital and labour is in the nature of economic development. That natural resources increased more rapidly than labour may be due to a specific conjunction of historical circumstances—the existence in the nineteenth century of large unsettled territories—and the situation may be different in the future. All that matters for the argument is that labour did in fact turn out to be the scarce factor and therefore a preoccupation in the early nineteenth century with labour-saving technology proved to yield higher returns than one biased towards saving capital or natural resources.

Compare two economies, one with continuing labour-scarcity and the other with, say, natural-resource scarcity leading into labour-scarcity as time goes on. Technical progress in the first will be biased in one direction—saving labour throughout the period. The second will have a natural-resource-saving bias, and will find it correspondingly difficult to switch to a labour-saving bias, which at a later stage is made necessary by labour-scarcity.

The assumption underlying this argument is that there are difficulties in taking over a technology. We have suggested that in a number of fields the Americans were led to invent machines so superior that they more than compensated for the high cost of American labour. If the machines were so markedly superior, it would, of course, have paid the English to introduce them, and had the English adopted the machines to the full extent warranted by their superiority, England would indeed in these fields have lagged behind the U.S.A. in mechanisation, but the lag would not have increased with time. The U.K. would have enjoyed the benefits of the new methods of which the U.S.A. had borne the costs of innovation.

There was of course a considerable exchange of ideas about industrial processes between the two countries. 'If our people invent anything' said an American witness before the Royal Commission on the Export of Machinery, 'England is the best market to bring it to, because they think you will have the greatest demand. If you invent anything, persons go to America and secure patents, in order to make use of it there'.[1] Processes of American origin were brought to England, by Englishmen who bought American patents, adapted them where necessary to English requirements and pushed their use in England; and later on the establishment of subsidiaries of American firms served a similar purpose. But there often seems to have been a considerable time-lag. Thus a number of the early American textile improvements took a long time to transplant. The first rotary machine for shearing the nap, invented in 1792 by an American, Samuel Dorr, was patented in England in 1794, 'but' writes the historian of the technology of the industry 'nothing more was heard of it here'.[2] Ring-spinning, invented in America in 1828, was introduced into England in 1834, 'but it soon disappeared and did not return until after 1850'; and it did not have much practical effect until the early 1870's.[3] Twenty or twenty-five years elapsed before Goulding's carding machine was introduced into England.[4] There was a long lag too in the introduction of American woodworking machines. In America, reported the Committee on Machinery, Blanchard's lathe 'has been used very extensively for about thirty years in the turning of shoe lasts, boot-trees, oars, spokes of wheels, busts, gun-stocks etc. . . . It is most remarkable that this valuable labour-saving machine should have been so much neglected in England, seeing that it is capable of being applied to many branches of manufactures'.[5] The delay in introducing the new American machine tools was the most significant lag. As we have seen, in one particular case—the manufacture of small arms—American machines and methods were introduced in the 1850's. But these methods, introduced on government initiative, failed to exercise any immediate effect outside

[1] P.P. 1854–5, L, Qu. 3038. See the discussion on pp. 115ff. above.
[2] *A History of Technology*, ed. C. Singer and others (Oxford, 1958), vol. IV, p. 305.
[3] *Ibid.* p. 291.
[4] Cole, *op. cit.* I, pp. 105–6.
[5] P.P. 1854–5, L, pp. 38, 39.

this narrow field. In the U.S.A., in the first half of the century the state arsenals had been the forcing houses of the new mechanical methods; they were responsible for many of the early technical advances and trained several of the engineers who later pioneered in other industries. The establishment at Enfield never played anything resembling this role. The manufacture of Browne and Sharpe milling-machines was not introduced into England until the 1870's, and the automatic turret-lathe until 1897, thirty years after the first such lathe had been made in America. It was not until the boom of the later 1890's that American-type machine tools began to be extensively manufactured in England and this boom also saw the adoption of American improvements in iron and steel.[1] From the 1850's to the 1890's the disparity of techniques tended to widen in America's favour.

No doubt one could explain a considerable lag in the adoption of American methods by reference to the usual inertias and frictions and it is usually to some failure of entrepreneurial ability, the inability of British manufacturers to see where their interests lay with sufficient promptness, that the lag is attributed. There was undoubtedly a tendency among the English in the early nineteenth century to underrate American achievements and to refuse to acknowledge that they had anything to teach. The English were very much less disposed to borrow from the Americans than the Americans from the English; the flow of ideas was heavily biased towards the west. This attitude lasted until the Exhibition of 1851, where American exhibits created something of a sensation. The lag is sometimes attributed to the fact that American labour-saving devices were often not suitable for the type or grade of product in which England's comparative advantage lay. This argument may sometimes have been simply a rationalisation of complacency and inertia, but there is no doubt that sometimes it was a

[1] For the American influence in the 1890's see S. B. Saul, 'The American Impact on British Industry, 1895–1914', *Business History*, III, No. 1, December 1960, pp. 21–7; D. L. Burn, *op. cit.* p. 185; J. H. Dunning, *American Investment in British Manufacturing Industry* (London, 1958). The British weekly *Engineering* in the 1890's contained symposia on American workshop methods, and the *Colliery Guardian* for 1901 had a series of articles on machinery in the American mining industry. The best introduction to the contemporary literature on American methods is R. H. Heindel, *The American Impact on Great Britain, 1898–1914* (Philadelphia, 1940), ch. IX.

genuine reason and not an excuse.[1] In a few cases also the terms on which raw materials were available may have made what was the best technique in the U.S.A. unsuitable in England. But besides these circumstances, there was the more general influence of abundant labour.

In the first place, there was the impediment to the introduction of new methods presented by the ability of the relevant domestic industry to cut its labour-costs in face of competition from the production of more highly mechanised methods. This was an obstacle to borrowings from America as much as to spontaneous development of new methods. At the least, this added to the difficulties of predicting the results of introducing a new method. American automatic machines for lock-making were introduced into England in 1851, but though they reduced the price of cheap lever-locks, they were not widely adopted, partly because of the extreme cheapness of handicraft work in the industry. For much the same sort of reason, the application of machines to the making of boots and clothes was deferred until the last decade of the nineteenth century. The abundance of labour in the out-working system must be the principal reason why nail-making, in the later nineteenth century in the Black Country was still predominantly a house industry, employed the 'Oliver' operated by the foot, while simultaneously in Pittsburgh the industry was highly mechanised and concentrated in factories.[2] But quite apart from these considerations, in devising and operating more mechanised methods the Americans acquired experience and skills which enabled them to operate identical machines at a lower cost in terms of output than would have been the case in England. Imported machinery is unlikely to be worked as skilfully as in the country of origin. 'The same ingenuity and watchfulness that cause it to be devised in one country cause it also to be worked to best advantage there'.[3] Moreover the active pursuit of more mechanised methods in many different fields yielded external economies in operation, that is the cost of getting replacements

[1] Thus the automatic looms were best suited to coarse standardised cotton sheeting, and for this reason were slow to be adopted not only in Britain, but also in those branches of the American industry which specialised in quality goods. Clapham, *op. cit.* III, 177; Jerome *op. cit.* p. 339.

[2] J. Schoenhof, *The Economy of High Wages* (1893), p. 225.

[3] Taussig, *op. cit.* pp. 25, 197–9, 251. The difficulties of adopting the techniques appropriate to a different factor-endowment may be seen in the case of wood-

and spare parts was lower in America. Individual new methods which, on the face of it, look as if they ought to have been adopted were surrounded by a penumbra of know-how which could only be assimilated after some delay. And a bias towards a particular kind of improvement takes a long time to correct itself because it gives rise to a particular entrepreneurial outlook which is eliminated only when people die.

The difficulties of taking over a technology are greater the more elaborate the technology. Thus it was easier for the Americans to borrow new English methods in cotton textiles in the early decades of the nineteenth century than it was for England to borrow American mass-production methods in the later part of the century. This is irrespective of the particular bias of the technology. But, in addition, it is probable that a technology biased towards the saving of labour could borrow more easily from a technology biased towards the saving of natural resources than vice versa, since labour-saving develops a high degree of the mechanical ingenuity which is necessary to put any invention into practice, no matter what resource it saves.

THE EXPANSION OF THE MARKET

Up to this point, in our discussion of the English experience, we have not paid attention to the rate at which the market for English goods was expanding. We have been principally concerned with the fact that, for a given rate of investment in relation to total factor-supplies, a shortage of one factor, labour, was more likely to appear in the U.S.A. than in England. We shall now consider the probability that, in relation to the total supply of factors, English investment was expanding less

working machines. In 1844 William Furness of Liverpool imported from the United States and patented in England a number of such machines. 'The ruling idea in these machines was economy in cost and rapid performance in the hands of skilled men, neither of which elements fitted them for the English market at that time; consequently no great use seems to have been made of the plans and modifications which they might have suggested to builders in England. During the Exhibition of 1851, however, the performance of the wood-framed American machines was such as to create astonishment. English engineers at once proceeded to clothe the "ideas" these machines suggested in a mechanism more in keeping with their purpose and the true principles of construction, and out of it grew ... a large share of modern practice in England'. Richards, *Treatise*, p. 17. For instances of steel processes which failed to transplant successfully see *Journal of the Iron and Steel Institute*, 1871, II, pp. 258, 271; 1872, I, pp. I–XXXVI.

rapidly than the American, so that the advance of English technology in the decades after 1815 was less sensitive to factor-endowment. We have already discussed this matter in connection with America. The point to be considered now is the limits on the rate of investment in England in the first half of the nineteenth century. Attention is normally concentrated on explaining why British economic progress during the industrial revolution was so rapid, compared with that in other countries. But perhaps the more interesting problem is why English progress in the first half of the nineteenth century, though very rapid, was not *more* rapid than it was. Labour was cheap and its supply more elastic than in the eighteenth century. Land, of course, was much scarcer in England than in the U.S.A. But in view of the course of agricultural prices and rents from 1815 to the late 1840's it cannot very plausibly be supposed to have been a brake in these years. Nor does it seem that raw materials acted as a check in this period either by temporarily stalling economic growth in the boom, as happened in 1857,[1] or by exerting a long-term inhibiting effect. The supplies of coal and iron were not unresponsive. There is no evidence of a bottle-neck in capital goods, and in any case full employment of capacity in the machine-producing industries is commonly held to be a stimulus to investment.

The most likely restraint on the factor side is finance. It is a commonplace that while the English banking system in the early nineteenth century was very well adapted to provide industry with working capital, it did not help to finance fixed capital. Banks invested their resources in liquid assets, bills of exchange, and shunned bricks and mortar. It is true that the more banking practice is examined, the more frequent are found to be the exceptions to this rule—exceptions made to accommodate relations and business associates—and in the aggregate the assistance so afforded may have been significant. Nevertheless the general view is probably correct that the principal source of fixed industrial capital was industral profits.[2] The debatable question is how far the dependence on industrial profits was due to their adequacy for the fixed in-

[1] J. R. T. Hughes, *Fluctuations in Trade, Industry, and Finance* (Oxford, 1960), p. 138.

[2] L. S. Presnall, *Country Banking in the Industrial Revolution* (Oxford, 1956), ch. 10.

vestment which industrialists wanted to undertake. It can be argued that financial institutions adapt themselves to meet the principal economic needs of their period and that English banks concentrated on the provision of working-capital because that was what industry needed; if there had been a large unsatisfied demand from industry to finance fixed capital, financial institutions would, with relative ease, have adapted themselves to meet this need, or new institutions would have arisen for the purpose, just as the German banks developed to afford long-term finance to industry in the 1850's. 'By and large, it seems to be that the same impulse within an economy which sets enterprise on foot makes owners of wealth venturesome, and when a strong impulse to invest is fettered by lack of finance devices are invented to release it . . . and habits and institutions are developed accordingly'.[1]

This is an argument the force of which in any particular case is exceedingly difficult to assess. For entrepreneurs frustrated by lack of finance, for that very reason, leave little trace. But there is no reason in principle why all financial institutions should make the adaptation called for, or why, even if they do, there should not be a considerable time-lag. As we argued in an earlier section, some financial institutions—those concerned with the national debt—developed in such a way that they competed with the claims of fixed capital for industrial profits. And the market in mortgages, which had originated to meet the needs of landowners and which still, in the early nineteenth century, principally served their needs, was difficult to adapt fully to the needs of small industrialists who might not always be able to offer security of the type normally required by lenders.

We certainly cannot argue from the fact that industry relied heavily on its own profits that these were adequate for the investment which entrepreneurs as a whole wished to undertake. In some cases they evidently were, for some industrialists invested heavily in government debt. But in other cases the rate of profit may have been insufficient to finance the investment which, at that rate, manufacturers wished to undertake. The cotton-textile industry was composed of a large number of small units facing a highly competitive market, and many of

[1] J. Robinson, *The Rate of Interest*, *op. cit.* p. 86.

the technical innovations in the industry, by the 1820's, had a short gestation period and yielded cost-reductions fairly rapidly. Increases in productivity, therefore, showed themselves as much (and possibly more) in lower prices as in higher profits; so that much of the advantage accrued to fixed-income receivers, a class which probably had a high propensity to save but which did not lend to industry. Some of the increasing productivity was, through taxation, siphoned off to pay interest on the government debt. It may well be therefore that not enough of the increased productivity accrued to profits to finance the investment plans which entrepreneurs were prepared to undertake. While in some industries the limiting factor on investment plans may have been the low rate of plough-back for socio-psychological reasons, in others even a complete plough-back would not have provided sufficient finance, and the difficulty of raising external capital may have been the effective limitation.

After the cessation of heavy war-demands in 1815, England had the appearance of a country of surplus capital. Until the Crimean war the British government did not borrow, and there were strong forces—particularly but not exclusively taxation—redistributing income in favour of those with an inclination to invest in government debt and mortgages. The rates on such securities were therefore low. But these rates are not a good guide to the terms on which capital was available to industry. Because of the preferences of monied men, and impediments to the flow of funds between industries and firms, and between industry and other employments, some manufacturers, especially the recently started, were possibly frustrated by shortage of finance at a time when others, especially the well-established and successful, had more money than they knew what to do with. Finance therefore cannot be ruled out as an effective limitation on growth. But it is not likely to have been a very serious one, and the impression remains that there was little on the side of factor-supplies to prevent English industrial capacity being built up between 1815 and 1850 more rapidly than in fact it was. What, then, were the limitations on the rate of growth?

For much of the eighteenth century the rate of investment in England had been rapid in relation to the increase in the supply of factors, including labour. Since the supply of factors was increasing even more rapidly in the first half of the nineteenth century it would have required an increasing rate, and a very large absolute increment of investment to have produced a very marked increase in the marginal costs of even the most sluggish factor. And a number of reasons can be suggested why English investment failed to increase at such a rate, quite apart from the fact that, in the most favourable circumstances, it takes time to organise investment.

One possible explanation is that English entrepreneurs were less single-minded. It was a common complaint in the late seventeenth century that the growth of English trade was checked by the withdrawal of mercantile capital and ability into landownership, since the aim of most merchants was to accumulate enough money to buy an estate.[1] Such motives still moved English manufacturers in the later eighteenth and early nineteenth centuries.

A manufacturer whose main aim was to make enough to cease to be a manufacturer would, for that reason, reinvest all the more vigorously until he had made enough, but when he had made his fortune not even a high return on further investment would induce him to continue, and the prospect of a fall might well precipitate his departure. Some of the entrepreneurs of the English industrial revolution did ultimately withdraw from business to found landed families.[2] The plant they had created of course remained, and that is on the credit side; and the loss of experience and ability was not great when they were old before they withdrew. Indeed where the plant was sold, it was to men whose ability to buy implies they could make good use of it, and possibly better use than the families which sold out; it was no doubt a loss to English industry that the second Robert Peel did not remain a mill-owner devoting his energies to textiles rather than to politics, but in other cases the sale may have mitigated the effects on management of

[1] J. Child, *Brief Observations Concerning Trade and Interest of Money* (1668).

[2] Disraeli is not evidence, but his view is interesting: 'From the days of Sir Robert Walpole to the present moment, with one solitary exception, all those who have realised large fortunes in our great seats of industry have deposited the results of their successful enterprise in the soil of their country' (*Hansard*, 3rd series, LXXXVI, 87, 86). I owe this reference to Jacob Viner. James Nasmyth retired from business when he was only forty-eight years old: 'my profits were so increased, that if I had wished to expand the establishment . . . if instead of putting it into three per cent consols I had put it into additional workshops, I might have taken on 400 or 500 or 1000 workmen' (P.P. 1867–8, XXXIX, Qu. 19139).

second and third generation industrialists. But even in the case of complete withdrawal from business, the sale of plant neutralised the effect on investment of the different sort of manufacturer whose main aim was to create a large concern. The possibility of expanding capacity by the purchase of existing plant set a limit to the rate at which those industrialists who were interested primarily in the growth of their concern would undertake fresh investment. The purchaser presumably got a better return than if he had bought new equipment, but the net result of the transaction was that the funds of the buyer were devoted to enabling the seller to buy land, that is to making it easier for English landowners to live beyond their means.

But more commonly the English manufacturer did not sell out entirely, but adopted the style of life appropriate to non-industrialists. Like Sir John Guest, they retained their works, but acquired estates and a London house, and diverted funds from reinvestment to maintaining a position in society. If their social expenditure was insensitive to the rate of profit on new investment, even a high rate would not induce them to invest; but even if they were prepared to tailor their social expenditure to the needs of the business, at low rates they might prefer a new town house to a new plant. The 'part-time' manufacturer probably had a much worse effect on the rate of investment than the man whose aim was to cease entirely to be a manufacturer, because he retained the ownership of the assets but did not completely devote either his energy or his funds to developing them. How frequent these types of manufacturer were in early nineteenth-century England must await investigation, but they were obviously more common in England than in America.

Lastly there is the possibility that the limit to the English rate of investment was set by the nature and size of the demand for English goods. The extent to which the incomes generated by investment in English factory industry could absorb the products of that investment at a rate of profit satisfactory to English capitalists was limited. There was, as Malthus asserted, a deficiency of demand. An increase in capacity and employment increased total expenditure on products by industrial wage-earners and by those in agriculture who provided them with food. But since additional labour could be obtained for agriculture at the subsistence-wage, and many types of labour could be obtained for industry at the subsistence-wage plus a small margin, the increase was probably not greatly in excess of the increase in employment. A large part of the increase in productivity went to fixed-income receivers whose demand for industrial goods was not elastic in respect of either price or income. The increase in industrial output depended, therefore, on effective demand in other sectors of the economy and in foreign countries. This depended partly on the increase of income of the customers of English factory industry and partly on the diversion of their demand from other suppliers of industrial goods. In England itself the demand for industrial goods was boosted by construction activity, for example the building boom of the 1820's and the railway booms of the 1830's and '40's. Abroad, income was increasing in North America, and a substantial part of the increase in demand spilled over on to English industry. But elsewhere the rise in incomes up to the beginning of the 1850's was probably very modest: sea-transport costs were still sufficiently high to hold up development in most of the other primary-producing regions, and not enough railways had been built in continental Europe to give much of a boost to income there. A large part of the demand for British industrial goods came therefore not from an increase of the income of its customers but from a diversion of demand, from enroachment on markets hitherto served by industry organised on the domestic system—both the English domestic industry and that of the Continent.

We were previously concerned with the effect of domestic industry on the composition of investment;[1] we are now concerned with its effect on the rate. The existence of domestic industries employing old labour-intensive techniques meant that up to a certain point demand for English factory products was very price-elastic. Once the initial difficulties of the new technology had been overcome and the substantial ancillary improvements made, the fall in the costs of English factory production assured a rapid increase in sales, and since transport-costs were falling, and market arrangements were improving, the geographical area over which they could compete with the local domestic industry increased. But for any given fall in manufacturing and transport-costs a limit was set to market

[1] See pp. 144–151 above.

expansion by the ability of the producers in domestic industry to tighten their belts and lower their prices. It may be suspected that in some industries this point was being approached in the 1840's and that this is one reason for the disappointing history of the decade. Domestic producers had cut their prices: hence the readiness of the workers in domestic industry, normally a conservative force, to undertake revolutionary action both in England and on the Continent. It is quite plausible to argue that the geography of the revolution on the Continent was influenced by the intensity of the competition of the English factory, and that the economic basis of revolution was weakest in those areas which were most protected by distance or tariff from English factory goods. It was Arkwright who was ultimately responsible for the revolutions of 1848! But these prices were so low that expansions of factory capacity were inhibited.

Thus, until the 1850's a limit was set to the rate of expansion consistent with a satisfactory rate of profit by the low incomes of English and continental consumers and by the price-inelasticity of supply of domestic producers. And it looks as if the rate so set was not high enough to run up against capital—let alone labour—shortage, in the sense that, had they been assured of better markets, English manufacturers could easily have found the finance for additional capacity.

Given the abundance of the supply of factors, English investment in the first half of the nineteenth century would have had to increase more rapidly than the American to run into sharply rising factor-costs. A rate fast enough to do this would almost certainly have first run up against inelasticity of demand for English products.

But the market was limited in part, because innovation was confined to a narrow range of industries whose progress was not shared by other, backward sectors. Why did not innovation influence a wider area in this period?

The answer lies partly in the economic characteristics of the largest innovating industry of this period—cotton textiles.[1] For reasons already suggested, the cotton-textile industry was in several ways well qualified to be an engine of growth. The demand for its products was price elastic; it produced standardised goods acceptable over a wide area not only in England

[1] P. Deane and H. J. Habakkuk, 'The Take-Off in Britain' (Papers of the International Economic Association Conference, 1960).

but abroad; the unit of production was small and consequently entry into the industry was easy; there was scope for cost-reducing improvements, and nothing to prevent these being reflected promptly in price reductions. The growth of the cotton-textile industry itself was therefore spectacular. But partly for the same reasons, the effect of its growth on the economy as a whole was restricted. The technical improvements made within the industry were not widely applicable outside. The multiplier effect of investment in the industry cannot have been great because its raw material was imported and its capital output ratio was low. And, just because the demand for its product was price-elastic, the fall in price did not release so much consumers' demand for the products of other industries. Moreover the cotton-growing states whose growth was stimulated by the increase in British demand for the raw material did not have a very high propensity to import British manufactures, or at least, were prevented by the United States tariff from indulging it fully. A large part of the demand for manufactures generated by the growth of income in the deep south was concentrated on American producers.

There are, it is true, considerations to be placed on the other side. The spectacular growth of the industry affected other sectors by force of example; it captured the imagination and fired the ambition of contemporaries. Even though its direct economic impact was restricted, entrepreneurs in other fields may have been readier to take risks and to innovate because of the success of cotton. It is not in any case the intention here to suggest that any other industry could have assumed the role of cotton textiles in this period. Nevertheless, to explain why England did not advance over a wider industrial front it is relevant to point out that, while the textile industry served a mass market and was therefore capable of growing to a considerable size, its interrelations with other industries were not of a kind which would automatically stimulate expansion elsewhere in British industry.

The extent of innovation in England was also influenced by the character of technical progress in this period. Technical progress has a logic of its own; one invention raises problems, intellectual and mechanical, which require time and effort for their solution and which, being solved, suggest new lines of advance. At any moment, only certain lines of advance are technically possible and with the best will in the world one can only go a certain distance. At some periods, for purely autonomous reasons, the technical knowledge available to be incorporated into practice is richer than at others. Compared with the last forty years of the eighteenth century the first half of the nineteenth was not a period of major technological break-throughs in English industry. Down to the mid-century and indeed well beyond, the principal advances sprang from remedying imperfections in the basic inventions of the industrial revolution and applying the principles involved in them to new fields. It was not until the 1870's and '80's that additions to fundamental knowledge assumed technological significance.

There are good reasons why technical progress within individual industries should have taken the form of break-throughs limited in time, rather than steady advance. The initial application of new techniques generally depended on a small number of dominant entrepreneurs who rarely perpetuated themselves and were often succeeded by men of routine; within an individual industry, therefore, a period of major innovation and expansion tapered off as control passed from the hands of the pioneers. Moreover, under the explosive expansionary influence of a major innovation, most of the relevant technical knowledge available at the time tended to be incorporated, so that, in the period immediately after, the remaining technical possibilities were limited to improvements and modifications. But even where the major innovation did not exhaust all the technical possibilities, and further new processes could have been invented, they were often not pursued; after a phase of innovation the most important firms were more concerned with preserving past gains and perfecting past inventions than with plunging into new technological ventures.

But why were there no major break-throughs in the industries which had not been leaders in the industrial revolution? Why did electricity and the internal combustion engine not become important earlier? The reason is partly that intractable technical problems had to be overcome before development was possible. Hence the long period between Faraday's discovery of the principle of electromagnetic induction in 1831 and the development of the dynamo in the later

1860's and early 1870's; and, again, the delay between the dynamo and the solution of the subsidiary problems which arose in the attempt to apply electrical power to new uses—problems of the long-distance transmission of power, the development of efficient lighting methods, the invention of the high-speed turbine and its application to ships. Likewise, in the development of the internal combustion engine to the point where it became a practical proposition, technical considerations were the dominant influence. No doubt one can imagine economic situations in which more and earlier attention would have been concentrated on these problems, resulting in earlier solutions; but the situations are rather far-fetched. The solution of the technical problems depended on ancillary developments in other industries, often made for quite independent reasons, and it is not likely that the whole process could have been much speeded up.

But, though there were few additions to fundamental technical knowledge, the improvement of basic processes, and the attempt to apply their principles to wider fields did yield considerable technical progress between 1815 and 1850. These improvements, though often uninteresting from a strictly technical point of view, were economically important and led to a substantial fall in costs and increase in capacity in the industries directly affected, particularly in cotton textiles. But the market set a limit to the rate at which this narrow range of industries could grow. Everything therefore depended on the extent to which these improvements could be made use of in other industries. In some cases they were not suitable for application far outside the industries in which they had been made; for example, the improvements in the techniques of the cotton-textile industry could only be applied to textiles, and even then within limits. Nor were improvements in the primary iron industry fruitful for technical advance in other industries. The technological improvements which had the widest possibilities of application were those in steam. Fellner has suggested that the application of steam to land-transport was limited in this period by technical considerations: the gap between the canal era and the railway era was due, he suggests, to the relative slowness with which the technological prerequisites of the railway became satisfied.[1] The same can be

[1] W. Fellner, *op. cit.* p. 99.

argued of the application of steam to sea-transport. The main technical problem was that the early marine engine was heavy in relation to its power and consumed a great deal of fuel. The principal solution for the high fuel consumption was compounding, that is the use of the expansive power of steam at different pressures in two or more cylinders, and the principle was patented by John Hornblower as early as 1781 and developed by John Elder in 1854–6. But it did not come into wide use because of the dangers of high-pressure steam in weak boilers. Developments in steel, particularly the improvement in the open-hearth process, were needed to make it practicable to build strong boilers and to apply the compound engine. This was just a happy coincidence of developments, for open-hearth steel was not the result of a search for stronger boilers, but stemmed from Siemens' patent of 1856 applying the regenerative principle to furnaces. In the application of steam to different types of machine and in the production of steam power at an attractive price there were also technical problems to be solved: and here too much depended upon independent progress in such things as the methods of machine building.

In a number of important cases therefore the technology available limited the pace of investment. This is not, of course, the whole story. For while technology influences investment, investment in turn enlarges the technical possibilities by affording opportunities to test new methods; whether a new invention is made depends on whether previous inventions have been made use of, because it is only the practical application of an idea which shows up its deficiencies, and indicates the promising points of future development. With a higher rate of investment the potentialities of the stock of technical ideas available in the early nineteenth century might have been more exhaustively explored. Thus if shipping had not been depressed after 1815 the problems of applying steam and new raw materials to water transport might have been tackled earlier. It is particularly difficult to believe that the application of steam to machinery could not in more favourable economic circumstances have been more rapid; for steam did not begin to play an important part in powering the British economy until the 1830's and '40's, and was not massively applied until the 1870's and '80's. Even as late as 1870 less than a million horse-power was generated by steam in the factories and workshops of Great Britain.

Nevertheless though technical potentialities were not always fully exploited, the nature of the technical improvements available does help to explain why English industrial investment did not grow faster. It also adds point to the fact that English engineering genius was not primarily devoted to developing labour-saving devices of a sort which might have had general application, and to the fact that, because labour was abundant, the proposals for uniformity in the dimensions and fitting of machinery, put forward by Whitworth, were very slow to be adopted.

We have been suggesting that much of the technical progress of the period took forms which were not capable of wide application outside the industries in which they originated. But even if the leading sectors had been more suitable for the purpose, innovation in one sector was less likely than in the eighteenth century to have wide repercussions because there was a good deal of slack in the economy. It is where resources are stretched that innovation is likely to reverberate. In early nineteenth-century England an innovating industry was less likely to be held up by shortages—of capital, labour and raw materials—but by the same token it was less likely to elicit or enforce changes on the rest of the economy.

What stimulated growth over a wide area in the quarter century before 1815 had been less the growth of the cotton-textile industry than (*a*) internal transport improvements especially canals and (*b*) rapid expansion of exports and re-exports. The canals widened the markets in products and factors. The fact that canal-building dried up about 1820, that is a decade or more before railway competition became widespread suggests that they had by then reached their economic limits; and their main impact had already been achieved. The increase in exports towards the end of the eighteenth century, though in part the result of cost-reducing innovations in English industry, owed more to the fortunes of war than to technological change. In the 1790's the French wars, by crippling Britain's major commercial competitors in Europe, opened up substantial new markets for domestic and entrepôt goods. Between 1790–4 and 1800–4 the volume of exports from Britain to Europe almost doubled, after several decades of stagnation, and those to North America increased by 58 per cent. To the extension of the market which took place in the 1790's and early 1800's the power of the British Navy contributed as much as the inventiveness of British industrialists. Some of the market gains were held after the war, but foreign demand for English goods was not again boosted by political events. Until the railways there was no other innovation which exercised an influence on the economy as a whole. And though exports continued to expand, this was largely due to the fall in costs of British cotton textiles which enabled them to be sold at the expense of competing products and producers. It was not until the 1850's and '60's that the British railways began to exert their full effect on industry. In the same decades the growth of the incomes of our foreign customers increased the demand for a wider range of British goods. Thus the years 1815–50 lie between periods in which internal transport improvements and vigorous demand for exports exerted a stimulating influence on a very wide range of economic activities.

We have been attempting to explain why market prospects in England for most of the first half of the century did not warrant a rate of investment sufficiently rapid to run up against severe scarcities of resources. We shall now consider the consequences of the fact that pressure on English profit-margins came more from the side of demand than from the side of factor-supplies. Why should pressure from this source be a weaker stimulus to technical progress?

Market difficulties did in fact lead to a search for improvements. Take, for example the English cotton-textile industry. For reasons already discussed, there was a rapid increase in output. When existing market opportunities were temporarily exhausted, factory owners did not cut production but tried to cut costs by increasing the scale of production in order to reduce overheads, and by improving their equipment. Provided the depression was not too severe, the reduction of profit-margins provided an effective incentive to make industry more efficient. 'Low profits', wrote Graham to Peel in 1843, 'stimulate ingenuity whereby machinery makes fresh inroads on the demand for manual labour'.[1] This was an incentive which also operated

[1] British Museum, Peel Papers. Robert Owen believed that the want of demand after 1815 stimulated the adoption of machines: 'every economy in producing was resorted to, and men being more expensive machines for producing than mechanical and chemical inventions and discoveries, so extensively brought into action during

in the U.S.A.; writing of American textile machinery Professor Navin observes that 'in times of depression a strong pressure to reduce costs always made itself felt on the mill floor where developmental ideas typically originated'.[1] But in so far as the market was more of a restraint in England than in the U.S.A., the incentive would have been more continuously felt in England.

There are, however, reasons why a given fall in the rate of profit on new investment caused by the inability of the market to absorb additional production except at much lower prices is less stimulating to innovations than a fall produced by the inelasticity of factor supplies. It is when the long-term market prospects are good that a manufacturer is likely to install new machines. Where market prospects are poor the incentive is, not to install new machines, but to reduce unit costs by making better use of existing machinery. Moreover the effects of inelastic factor-supplies becomes most evident during the trade-cycle boom; the effects of long-term inelasticity of demand during a slump. A boom emphasises a long-term in elasticity of factor-supplies and disguises a long-term weakness in demand; a slump has the reverse effect. The capacity created during the boom may reflect decisions taken during the previous slump and only put into effect when boom profits permit, but it is likely to reflect to a greater degree the conditions existing when it is created. Since most capacity is created in booms it is factor-shortages which have most effect on the composition of investment.

the war, then men were discharged and the machines were made to supersede them' (*op. cit.* p. 172). The reports of the factory inspectors give details of the reaction of cotton manufacturers to depression. Particularly in the years 1837–41 they attempted to reduce cost of production by both capital- and labour-saving methods: 'by rendering existing machinery more productive; by superseding manual labour by mechanical contrivances and where manual skill is still necessary, by getting it performed by children and young persons instead of adults'. In particular improvements were said to have been made in the operation of mules, by making them carry a greater number of spindles, by making one man work four instead of two, and by the introduction of self-acting mules. P.P. XXII, p. 362. 'When bad times come upon him [the employer] it drives him to consider where he can economise' [James Nasmyth, R.C. on Trade Unions, 1867–8, XXXIX, Qu. 19155]. W. W. Rostow argues that in the 1870's the rigidity of wages in relation to prices stimulated the invention of labour-saving devices. *The British Economy in the Nineteenth Century* (Oxford, 1948), p. 75.

[1] Navin, *op. cit.* pp. 113–23.

Thus the economic contrast between America and Britain in the early decades of the nineteenth century is not simply that industrial labour was dear and difficult to obtain in the former and cheap and abundant in the latter. This was a point of difference, and it is capable of explaining why American entrepreneurs should have sought to save labour by the substitution of machines. But it is of at least equal importance that American investment was more rapid than the English in relation to the supply of all factors except agricultural land, so that the American entrepreneur was more likely to be subject to constraints imposed by factor-limitations. If despite these constraints he wished to expand capacity his choice of technique would be sensitive to factor-proportions and he would be under pressure to seek for methods which economised capital as well as labour, though labour would remain his main preoccupation. He would wish to install new capacity not only because of autonomous technical progress but because of the technical progress accruing from attempts to offset factor-constraints and because of the increased demand for industrial goods as the country was opened up. This process of opening-up took the form of a series of development booms which had conflicting effects upon industrial activity, on the one hand, competing for labour and possibly capital, but, on the other hand, enlarging the market for industrial goods. It was the interplay between constraints on the factor side and the prospects of expanding markets, the combination of frost and sunshine in the economic climate, which was the distinguishing feature of American development.

In the case of Britain in the same period, though the absolute amount of new capacity created in those industries which were experiencing technical progress was large enough to afford entrepreneurs plenty of opportunities to try out new techniques and acquire experience, it was not large enough in normal circumstances to run up against significant restraints from the side of factor-supplies. When the British entrepreneurs created new capacity they did often incorporate a great deal of new technology devised from the experience of the defects and possibilities of existing methods. But there was no pressing reason why they should change techniques unless the superiority of the new was evidently great; they were not under the same spur as the Americans to try anything that looked the least promising.

VI

TECHNOLOGY AND GROWTH IN BRITAIN IN THE LATER NINETEENTH CENTURY

In the first half of the century, the problem was to explain why America's technical progress should have been more rapid than Britain's in certain industries. The industries concerned were few, and the problem attracts attention only because it is, on the face of it, surprising that even in a small number of industries a country so predominantly agricultural should have been superior. The absolute contributions of Britain to technological progress in the period must have been much greater.

From the 1870's on the problem is a more general one. What needs to be explained is the widely held impression that in the forty years or so before 1914, advance in technology over industry as a whole was slower in Britain than in the U.S.A., and also than in Germany; her own contributions were fewer, and technical developments of foreign origin were less widely and less promptly adopted in Britain than British developments were adopted elsewhere. There was a technological lag in the most important of the basic industries: the efficiency of British iron and steel plants was thought to be far behind that of their competitors in the U.S.A. and Germany. And there was also a lag in the new industries which were becoming prominent in the closing decades of the century, as shown by the excess of English imports over her exports of cars, electrical equipment and certain types of chemical.

The extent to which the U.K. was falling behind by relevant economic criteria has still to be established, but some lag in technology is reasonably well attested by contemporary comment and requires explanation.[1] There is a widespread impression that this lag represents the onset of deep-seated and persistent deficiencies. The changes are frequently presented in terms of rigidity and ossification, the characteristics of an old economy,

[1] For a balanced account of Britain's technological performance from the 1880's on see J. H. Clapham, *op. cit.* III, ch. III.

and when the explanation is not anthropomorphic it is sociological in character. The form which the sociological explanation most often takes is to attribute these technical deficiencies wholly or in part to British entrepreneurs. This explanation is applied to all the British basic industries but most conspicuously to the heavy industries.

There is a general argument, to which we have referred in a previous section, that the English social structure and English public opinion were less favourable than the American to entrepreneurship, less favourable both to the recruitment of ability and to full exertion of ability once recruited.[1]

In the United States, there was no long-established class system to impede social mobility. The possibility of rising to the top was believed to be greater than anywhere else in the world, and probably was so in fact. This belief in an open avenue to wealth was one of the main reasons for the amount of ability devoted to entrepreneurship in the United States. Moreover, in the United States, there were few competitors to business success as a source of social prestige. There was no large and powerful bureaucracy, no hereditary aristocracy. There was no professional military class and soldiers were not held in high esteem. Horatio Alger, the hero of the American success story, wanted to be a business man, not a general or a civil servant or a great landowner. The men of ambition and ability turned naturally to business, not only because of the gains which might be made there—though they were sometimes certainly enormous—but because business men, a Rockefeller or a Pierrepont Morgan, were the leading men of the country. Morgan himself, in his early middle age, nearly decided to give up business and become a gentleman; but he did not do so, and later American business men did not even consider the possibility.

In England, as, of course, in other countries of Western Europe conditions were more complicated. There existed a strongly entrenched social system which limited social mobility, a social system inherited from pre-industrial times when landowners were the ruling groups. Moreover, there were sources of power and prestige besides business. Landownership, bureaucracy, the army and the professions were all powerful competitors of business for the services of the able men. There was therefore a haemorrhage of capital and ability from

[1] See pp. 111–15 above.

industry and trade into landownership and politics. Robert Peel and Gladstone both came of entrepreneurial families, but their abilities were devoted to politics not to industry and trade.

In particular the high social standing of the professions drew off a large number of the ablest men. One of the most successful British entrepreneurs of the twentieth century, Lord Nuffield, is said to have entered business only because his father was unable, for economic reasons, to fulfil his original intention of making him a surgeon.[1] In England in the first forty years or so of the century, the professions attracted too many able people and business too few. After 1902, there was a vast increase in educational opportunities in England, but of those who profited from such opportunities up to 1939 many more became teachers or lawyers, doctors or architects than business men. Why were the professions more attractive? Certainly not because they yielded higher incomes; it was partly because, for a person with no personal contacts in business, the professions were more accessible, and partly because business in general lacked the social prestige of the professions. Moreover in the present century a business career and the acquisition of profit became actively disapproved of among the many people who were influenced by socialist ideas about capitalism and the profit motive. In England, therefore, as in Western Europe generally, business has had to face greater competition for the able men than it has had to face in the U.S.A. The wider the circle from which a country draws its business men, the more likely it is to produce great entrepreneurs.

Moreover, much the same circumstances which facilitated the recruitment of entrepreneurs in America were also favourable to the performance of men once they had become entrepreneurs. In the United States, almost from the beginning, birth and hereditary status were of small influence in determining a man's social standing as compared with the highly efficient performance of a specific and limited function. There the man who concentrated his energies on a single goal and achieved success in his occupation was the man who commanded respect; the accepted ideal was that he should rise as far in terms of wealth as his abilities could carry him. A man's

[1] P. W. S. Andrews and E. Brunner, *The Life of Lord Nuffield* (Oxford, 1955), p. 37.

success in his occupation was not invariably judged by the amount of money he made. Some occupations were very highly regarded which did not yield very high rewards, for example, membership of the Supreme Court, but income was certainly an important criterion of occupational success. The signs of social standing could more easily be acquired with money. In England, on the other hand, birth and family, education, behaviour, manners, accent, played the dominant part in determining social standing. The ideal of the 'all-round man', the man of well-developed but non-professional competence in many fields, the ideal of Renaissance Europe, was still strong; and, within the limits left by birth and family, social standing was influenced by the whole range of a man's achievements and abilities, his physical strength or weakness, the way in which he ordered his life, his capacity to paint, or to play instruments, his whole character and personality. A society in which the main emphasis is placed on success in one's job as measured by income is more favourable to the full exercise of business abilities, and for this reason England was less favourably placed than the U.S.A.[1]

These are general characteristics of English society which might militate against English entrepreneurship (and even more against that of continental European countries) at all times. But there are additional reasons which apply with particular force to the later nineteenth century. There are two assertions which need to be distinguished: the first is that English entrepreneurs were apathetic, by comparison with their American and German contemporaries, that is were men of less general capacity and force: the second assertion is that, though still able, they were men of abilities which had ceased to be relevant. Because England industrialised earlier, English entrepreneurs were more likely to be second- and third-generation entrepreneurs; as such they had less need and incentive to exert their full effort, and were more likely to be distracted from concentration on business by the possibilities of social life, which are open to the man who has arrived but not to those who are still climbing. The drive inside the individual entrepreneur to expand his concern, to make the most of

[1] For a discussion of this problem see F. X. Sutton, 'Achievement Norms and the Motivations of Entrepreneurs' in *Entrepreneurship and Economic Growth* (Harvard University Research Center in Entrepreneurial History, 1955).

opportunities, is greatest in the early stages of industrialisation and loses some of its force once an industrial society has been created. This, it is argued, is true whatever the form of industrial organisation, but it applies with most force to the members of a family concern, which was probably still the principal form of organisation at this time: the impetus in the first generation is very great but it loses force with succeeding generations. Second- and third-generation business men tend to be less energetic (an argument, however, which is rarely applied to second- and third-generation Professors).

The second assertion is that, quite apart from any decline in the quality of the entrepreneurs, their virtues were not the relevant ones in the later nineteenth century. In some industries the course of technical and economic change called for the exercise of qualities different from those by which the entrepreneurs had established themselves. In the iron and steel industry, for example, the founders of the older British firms succeeded by the operation of qualities which were not those most required in the 1880's and '90's. The ability most evidently absent was technical expertise. The later innovations embodied more technical knowledge than the early ones, that is there were fewer within the scope of the ingenious artisan. Furthermore, many more of the later technical innovations consisted in applying the techniques of one industry to another, in the cross-fertilisation of the techniques of different industries. In such cases even an entrepreneur who had acquired from experience considerable technical mastery, was at a disadvantage compared with someone whose technical training had been more formal and so more general. Altogether therefore British industrialists tended to be less well equipped than their rivals, 'to judge the commercial prospects of innovations while these were in their experimental change' and 'to forecast the trend of technical change'. The English employer was, contemporaries said, more of a commercial man and took less interest in the technical part of the work than those of France and Germany. There was, said Siemens, 'more prejudice against innovations'.[1]

That in late nineteenth-century England there were entrepreneurial deficiencies of a social origin few students of the period would be disposed to deny, but exactly how important

[1] D. L. Burn, *op. cit.* pp. 296–306, has a discussion of the personal factors.

they were it is very difficult to say. There is no reason to believe that the English business man, though he may have been less well equipped technically than the German, was less well-qualified than the American. Moreover, in general the argument ignores, or at best does not explain, the curious patchiness of English business performance in this period. The rapidity of technical advance in shipbuilding and in the open-hearth sector of the steel industry, for instance, show that the second generation of entrepreneurs in family firms could be conspicuously successful. And many of the deficiencies can be explained in more narrowly economic terms and with less recourse to sociological influences.

Part of the contrast in technology can be accounted for by differences in factor-endowment on lines similar to those already discussed in earlier sections.

In the later decades of the century, it is true, the course of technical change was less subject to this influence. Scientific knowledge became more important as a source of invention, compared with empirical, artisan trial and error, than it had been earlier; and though scientific invention is, of course, often guided by the need to cope with particular bottle-necks, the scientist, even when searching for new methods as deliberately as the artisan or professional inventor, is likely to be more sensitive to the technical possibilities of particular lines of investigation, and to give them priority in his search. Since science is also pursued for its own sake, the autonomous accumulation of scientific knowledge is much more likely to throw off accidental inventions which are not tailored to fit the needs of particular factor-endowments. For the same reason, scientific invention is more likely to produce sudden jumps, discontinuities in the spectrum of existing techniques, which reduce the scope of the influence of factor-supplies on the choice of techniques.

Moreover, there was probably less difference in this period between the labour supplies available in the U.S.A. and in England. The stream of immigrants into the U.S.A. supplemented the supplies of native-born labour; and, with the increase in the size of the industrial sector and improvements in transport and in the labour-market, industrial labour became much more readily available. And if American labour was, except in the remoter parts of the country, no longer scarce, in

England it was no longer as abundant as it had been earlier in the century. The fact that industrial real wages were rising in the middle decades of the century is perhaps evidence that the demand for labour was out-running the expansion of the labour force. In the peak of the 1868–73 cycle, there was a very rapid rise of wages, and this is the first boom of which it might be argued that labour shortage acted as a ceiling.[1] The rise of real wages of English agricultural labour in the 1880's and '90's certainly suggests that the surplus of agricultural labour had been absorbed in English industry or by emigration.

Not only was English labour less abundant than it had been earlier. In the middle decades of the century there was a rise in its productivity as a result of technical progress, capital investment and improvements in quality. Commentators sometimes explained differences between England and continental Europe in terms of dear English labour. Agricultural tasks which on the Continent were performed by labour—the controlling of cattle and the carriage of crops—were in England imposed on fences in the first case and on horses in the second.[2] Contrasting the manufacture of guns in England and Belgium the chairman of the Birmingham Small Arms Company wrote in 1866 that the English had an advantage over the Belgians in that 'more extended use of machinery, the use of which in Liège is discouraged by the cheap rate at which hand labour can be maintained'.[3] 'The cheap labour at the command of our competitors' thought Brassey, 'seems to exercise the same enervating influence as the delights of Caphua on the soldiers of Hannibal'.[4]

All these were significant developments. But on the other hand though science became a more important source of technology in the later nineteenth century, both American and British technology remained predominantly pragmatic until

[1] The difficulties of obtaining labour for puddling were so great that something like the American labour situation was produced. 'It would be to the interest of iron manufacturers' said one of their number 'to adopt a machine (for puddling) even if the cost of it should be greater than that of hand-labour. It seemed, indeed, that unless mechanical puddling could be achieved, they would not be able to get the necessary puddling done at all' (*Journal of the Iron and Steel Institute*, 1871, II, p. 268).

[2] Nassau Senior, *Industrial Efficiency and Social Economy*, ed. S. L. Levy, I, p. 218.

[3] Timmins, *op. cit.* p. 396.

[4] Thomas Brassey, *Work and Wages* (1872), p. 142. Chapter V of this work, 'Dear Labour Stimulates Invention', contains several instances of its theme.

well into the twentieth century. Over most of industry there was still a range of choice sensitive to relative factor-prices.

Moreover there were still important differences in labour supplies between Britain and the U.K. It is easy to exaggerate the change in conditions in Britain; between the boom of 1869–73 and the height of the 1914–18 war there was never a general shortage of labour. The disappearance of the rural surplus was to some extent misleading, for much of it had merely been shifted as surplus labour to the towns, in response perhaps to the building boom of the 1870's. 'Has there ever', asked one observer in 1914 'in the big towns at least been a time when employers could not get practically at a moment's notice all the labourers they required?'[1]

More to the present point, the cost of labour continued to be higher in the U.S.A. than in England. Though returns in agriculture had ceased to act as a floor to American industrial wages except in the far west, this function had been taken over by those sectors of industry which had most successfully adopted labour-saving equipment, and made technical progress. The rise in productivity of American industrial labour (more rapid than the English) reflected itself in the terms on which labour was available to new industries and undertakings; it meant that unless these could pass high labour-costs on to the consumer, they had to adopt the techniques which would raise the productivity of their labour sufficiently to enable them to pay the high American wages.[2]

There were moreover marked regional differences. The extreme imperfections of the labour- and product-markets had disappeared, but the labour-market was still probably less

[1] W. Beveridge, *Unemployment* (1930), p. 62.

[2] Brassey (*op. cit.* p. 136) estimated that wages were twice as high in America as in England. B. Weber, 'The Rate of Economic Growth in the U.S.A. 1869–1939', *Oxford Economic Papers*, v (June 1954). Rising productivity, without a commensurate increase in money wages, lowered the cost of labour in the industry which enjoyed the rising productivity. But it raised the cost of labour to industries where productivity had not been increased. The effect on a new industry can be seen in the case of tin-plate. Attempts to establish this very labour-intensive industry in the U.S.A. were not successful until the 1890's and then only under the protection of the McKinley and Dingley tariffs and partly as a result of the previous emigration of skilled Welsh workers in the depressions of the preceding period. Once the industry was established, the American makers sought to improve the efficiency of the pack-mill equipment, their improvements later being adopted in South Wales. W. E. Minchinton, *The British Tinplate Industry* (Oxford, 1957), p. 69.

perfect than the product-markets. Wages were higher in the mid-west than in the east and the labour-supply was less abundant,[1] so that it paid mid-western manufacturers as they expanded to import machines made with the cheaper labour of manufacturers further east. Some parts of the story we have told about America and England in the first half of the century can be retold for the mid-west and east of America in the second half, with the difference, of course, that there was no internal tariff.

For these reasons American manufacturers were still under a more compelling need than the English to raise the productivity of their labour. The principal method of doing this was to equip the labour with superior machines, and by the later decades of the century these were generally cheaper, in relation to simpler machines, in America than in Britain. In some types of machine, it is true, the Americans were markedly inferior to the English. At the end of the century the machinery for preparation, spinning and weaving of cotton cost about 50 or 60 per cent more than in Lancashire; and carding engines, cotton combers, drawing frames and other kinds of English textile machinery found a market in the United States despite a 45 per cent tariff.[2] Even in textile machinery, however, the disparity of costs between the two countries was smaller than it had been in the 1830's and '40's. And probably over the whole engineering field, machines in America were cheaper in relation to labour, and in some cases absolutely cheaper as well. This provided an additional reason for the American entrepreneur to replace labour by machines to a greater extent than did his English counterpart.

But the entrepreneur's choice of techniques was influenced by inherited attitudes as well as by contemporary labour conditions. The direction in which technical progress is made depends on the type of problem that inventors and manufacturers are most alive to, and the sort of ways of tackling it which are thought appropriate. Americans had acquired the habit of concentrating on those problems which lent themselves to a solution by the invention of labour-saving equipment, and on labour-saving (as opposed to other factor-saving) solutions to common problems. For this reason it is quite

[1] Report on Cost of Living, American Towns, Cmd. 5609, 1911.

[2] Young, *op. cit.* p. 8.

possible that invention in the second half of the century was more responsive to a given degree of labour-scarcity than it had been in the first.

In England on the other hand both employers and workers were still conditioned by attitudes engendered by abundant labour, and reacted to problems in mechanisation in ways that cannot be fully explained by contemporary factor-proportions. The workers still feared the introduction of more mechanical methods as a threat to their jobs. Disputes no longer arose, as they had done earlier in the century, on whether machinery should be introduced, but centered on the conditions of its introduction.[1] In the 1840's and '50's workers in the engineering industry had opposed the introduction of planers and lathes; in the 1890's they attempted to impose conditions on the use of the capstan, turret lathe, miller and borer which rendered the profitability of their introduction uncertain.[2] When employers in the 1890's extended the use of machines in the manufacture of boots and shoes, the workers attempted to maintain the cost of production by machines on a par with the cost of hand methods.[3] Thus the attitudes of labour still imposed a restriction on the choice of technique in England which was not present to the same extent in the U.S.A.

But for his part the English employer too had traits acquired from long experience of cheap labour. As a spokesman of the English boot and shoe manufacturers said in 1901: 'For years labour (in England) has been so cheap and has been content to work under such conditions as to render it a matter of small importance as to the mechanical assistance with which it should be furnished. ... Men have been cheaper than machines. Today ... men are getting dear and machines are getting cheap. The whip of dear labour was applied to the backs of American manufacturers years ago'.[4] Because English employers had been so long accustomed to cheap labour their inclination was apt to be to get more work at the same money wage. 'Nothing is more frequent', wrote one group of American observers of British industry, 'than the remark that the

[1] Webb, *op. cit.* p. 295.

[2] James B. Jeffreys, *The Story of the Engineers, 1800–1945* (1946), pp. 35–42, 142.

[3] A. Fox, *A History of the National Union of Boot and Shoe Operatives, 1874–1957* (Oxford, 1958), pp. 206—7.

[4] John Day, *Shoe and Leather Record*, 27 March, 1891, quoted in Fox, *op. cit.* p. 206.

working-man does not need more than so many shillings a week. . . . This view among employers has prevailed for so long and is so nearly universal that their every effort is to obtain more work for a traditional wage rather than to decrease the cost of production by means which will justify a higher wage. . . . Working-men have come to accept the view widely too and it is the acceptance of this theory of status which is at the bottom of the deadlock in British industry'.[1]

Even when labour-saving devices were introduced, many English employers were so habituated to the idea of low money wages that they were not prepared to concede to their labour the higher money earnings which the new devices warranted and which would have reconciled labour to their introduction. When steam was introduced into the Coventry ribbon trade in the 1850's, and machinery into the making of boots and shoes in the 1890's, the difficulties which arose were partly due to the unwillingness of employers to 'view with equanimity the prospect of paying their workmen any larger amount per week than that to which they are accustomed'.[2] American manufacturers by contrast, compelled from the start by dearness of labour to mechanise, were less concerned with money- than with efficiency-wages.

It was not only the attitudes of workers and employers which were influenced by abundant labour. In the 1850's a leading English engineer commended the use of reaping machines in England *not* because they saved labour, but because they enabled the farmer to secure his crops in the worst of seasons. 'In a variable climate . . . where a whole harvest may be lost or seriously damaged unless rapidly cut . . . the machine reaper becomes invaluable'.[3] Herman Merivale, a leading English economist wrote in a memorandum for the Commission on Trade Unions that 'properly considered, piece work is strictly analogous to machinery. It is a mode of obtaining the execution of more work, or better work, at less cost to employers than by the ordinary method of daily wages'.[4] Whatever the formal correctness of this observation an American would have been quicker to see the difference between making the labourer work harder and giving him better tools.

[1] Quoted by Saul, *op. cit.* p. 27.
[2] Webb, *op. cit.* p. 400.
[3] W. Fairbairn, *Useful Information for Engineers*, Third series (1886), p. 202.
[4] P.P. 1868–9, xxxi, p. 357.

How far the choice of techniques was the product of contemporary cost-conditions and how far of inherited attitudes moulded by the cost-conditions of the past it is impossible to say, but both influences were certainly at work in the English coal and steel industries and in some measure explain the technical differences between the English and American industries. In the coal industry in 1913 more than 40 per cent of the American coal output was mechanically cut, as against only 8½ per cent of the British. The reasons for this disparity were partly geological: conditions in British mines were less favourable to mechanisation. Something, moreover, was due to deficiencies in the British electricity industry and something, no doubt, to entrepreneurial deficiencies. But a major obstacle was certainly that the profitability of the improvements which were available after 1880—the coal-cutter and the conveyer—was very sensitive to the cost of labour, and that in England the saving of labour did not appear to warrant the costs of the machines. 'Generally speaking', said the official spokesmen of the mine-owners before the Samuel Commission in 1925, 'the saving in labour charges at the face is absorbed by the capital charges on, and the running costs of, the machine. . . .' Whether this view was correct it is quite impossible to say. But what is clear is that sufficient labour was available on terms which deprived mine-owners of the need to put the matter to the test. Between 1883 and 1913 there was a very large increase in the labour-force in the industry, accompanied it is true by a rise in wage rates, but by a rise which was easily passed on to consumers; indeed, as Mr A. J. Taylor says, 'it was advancing coal prices which carried mining wages upwards rather than *vice-versa*'. 'There seems . . . to have been no lack of recruits coming forward [from the land and from the large national pool of unskilled workers] at wage rates corresponding to prices which the consumer was already offering for coal'. Even at the height of the boom the coal industry was not 'handicapped by a shortage of labour at wage rates which it was both able and, in the last resort, prepared to pay'.[1] Moreover, because the supply of labour was abundant in the long run, the workers did not feel sufficiently secure of their prospects to co-operate during the booms, when demand for labour was

[1] A. J. Taylor, 'Labour Productivity and Technological Innovation in the British Coal Industry, 1850–1914', *Econ. Hist. Rev.*, vol. XIV, No. 2 (August 1961).

strongest and therefore the incentive to the mine-owner to mechanise was greatest; and since mechanisation in this industry, more than in most, involved considerable reorganisation of working schedules, it could not easily be carried out in face of worker's opposition.

In the steel industry, the improvements in American technique were, initially at least, primarily labour-saving. When the Bessemer plant, a British development, was introduced into America, American improvements took the form we have noticed earlier, that is labour-saving equipment and measures designed to speed up the work and increase output per unit of capital. Cheap labour cannot be the main reason why British makers failed to make and were slow to adopt the improvements made in America. For Germany and Belgium, where labour-costs were lower, adopted labour-saving devices more widely in the 1890's than did Britain.[1] But it must have reduced the effort which British steel-makers devoted to the matter. And the attitudes of workers, which were in part a product of the fear of unemployment, were also unfavourable to mechanisation.[2]

Moreover the fact that manufacture with interchangeable parts was a practice of long standing in America gave her an advantage in the production, not only of machines, but of complicated mechanisms of most kinds. As a result of long preoccupation with labour-saving methods Americans had developed types of engineering skill which were a prerequisite for the development of new methods wherever they arose and whatever factor they saved. At the end of the century, mechanical engineering was often still sufficient by itself to develop a new idea, and even where it was not sufficient it was still generally essential. Many of the ideas came from abroad. 'Until well into the twentieth century' it has been said, 'Americans have been content to let most of the basic discoveries in science and technology originate in Europe, while they themselves have followed a policy of adapt, improve and

[1] D. L. Burn, *op. cit.* pp. 47, 172ff, 184.

[2] The American Journal *Iron Age* in 1874 put forward the case that British ironmasters had enjoyed pauper labour and had for that reason not found it economically desirable to afford 'encouragement to inventive talent to devise labour-saving machinery'. (Quoted, in order to deny it, by Lothian Bell, 'Notes of a Visit to Coal and Iron Mines and Iron Works in the United States', *Journal of the Iron and Steel Institute*, 1875, p. 139.)

apply'.[1] This was particularly the case with inventions which resulted from the application of advanced scientific principles. The central idea behind an invention of this kind was easily borrowed, and where its further development into a practical commercial proposition depended upon mechanical ingenuity, it might be developed more successfully in the U.S.A. than elsewhere. In electricity, for example, the early dynamo improvements which preceded the efforts of Edison and Bush were the work of Gramme, a Frenchman, and Von Hefner Alteneck, a German. In electric traction the important initial steps were taken by Julien, a Belgian. Garz and Company of Budapest were the first firm to perfect the transformer, and it was from them that Westinghouse got his ideas. Many American inventors in this field were Europeans born and trained. E. Weston, an important figure in early electric lighting and C. P. Steinmetz, one of the best electrical engineers at the end of the nineteenth century, went to the U.S.A. in their twenties. But, though the big inventions originated mainly in Europe, their technological development was carried furthest in the States.

The internal combustion engine was another borrowed idea which was most successfully developed by America. There is no mystery about the rapid growth of this industry in a country of great distances and high incomes. More illuminating for the present purpose are the difficulties of the early development of the industry in England, for they show how ill-adapted were the craft traditions of English engineering to the mass production of motors-cars. The first petrol-driven car was produced by Daimler in 1887, but it was not until 1896 that motor-cars were assembled in Britain, and not until 1903 that they were completely built here. The necessary mechanical knowledge was widely spread in the English general engineering industry and this was the obvious source of enterprise, but general engineering firms were slow to enter the field. Possibly the domestic boom at the end of the 1890's so fully occupied the industry that there was little incentive to branch out into anything else; several engineering firms, it is true, did go into electricity in these years, but they did it rather as a side show and got out again when prospects clouded. But in the 1880's there must have been plenty of slack in the industry, and the

[1] J. B. Rae, 'The "Know-How" Tradition: Technology in American History', *Technology and Culture*, vol. 1, No. 2 (1960), p. 141.

failure to exploit the possibilities of the internal combustion engine at this stage perhaps reflects a lack of initiative. Furthermore British engineering firms did not turn to making motorcar parts as readily as American machine shops: the earliest English pioneers in the motor-car industry were handicapped by the need to make a high proportion of their own parts. In 1899 when F. W. Lanchester started to manufacture cars 'no ancillary trades had then developed and we had to do everything ourselves, chassis, magnets, wheels, bodywork, everything except the tyres'. Lanchester's experience isi lluminating because he decided to manufacture the components of his car on a system of interchangeable parts. His insight into what was needed had a touch of genius, but his idea was not practicable in British circumstances. He had, for example, to produce his own range of standard threads, and the individualistic craftsmen upon whom he had to draw were reluctant to work to standard instructions. 'In those days' he wrote later, 'when a body-builder was asked to work to drawings, gauges or templates, he gave a sullen look such as one might expect from a Royal Academician if asked to colour an engineering drawing'.[1] The attempt by an individual to create independently a system of interchangeable parts proved, in the short run at least, intolerably expensive. The limitations of English methods showed themselves again later when William Morris decided to buy his main components from the specialist makers instead of making them himself. The specialists proved unable to take orders as large as he wished to place, and 'could not, or would not, at this time (1913) meet him by expanding'.[2] Morris therefore turned to America for components.

Thus the technology and attitudes shaped by the experience of dear labour in the past, as well as dear labour at the time, account for some of the difference in technical progress in Britain and America in the later nineteenth century. But after 1870 there was another influence—the amount of new capacity created. In this period the absolute volume of new investment was greater in America than in England for reasons

[1] P. W. Kingsford, *F. W. Lanchester* (1960), pp. 40, 47–8.

[2] Andrews and Brunner, *op. cit.* pp. 59–60, 71. For the American experience see J. B. Rae, *American Automobile Manufacturers: the First Forty Years* (New York, 1959).

which, initially at any rate, had little if anything to do with the relative efficiency of the industries of the two countries; because more capacity was created, the Americans had more opportunity to incorporate new technical knowledge, to acquire know-how and so to make further advances.

To clarify the argument it is necessary at this point to consider the long-term changes in the market for Britain's industrial products. Britain was the first country to industrialise and for a surprisingly long time she remained, in absolute volume of production, the leading industrial country. Even as late as 1870 the United Kingdom still had 31.8 per cent of the world's manufacturing capacity, compared with 23.3 per cent for the U.S.A. and 13.2 per cent for Germany.[1]

That Britain's lead was sustained for so long a period was partly due to institutional features in other countries which impeded their attempts to take over the British technology. But, quite apart from this, England's early start gave her certain continuing advantages. Because the British made the first important improvements in cotton textiles, in the primary iron industry and in steam, they were in a favourable position to make the subsequent improvements. In particular, they were able to build their railway system before other countries: demand conditions were more favourable—railway building in England merely involved joining up existing centres of population and industry, not building ahead of demand as in continental Europe and the U.S.A.; and capital for railway building was more plentiful. Britain was therefore the first to enjoy the great external economies of a railway system; she was also in a favourable position to supply other countries with their railway equipment when they started to build.

In an earlier chapter we considered the possibility that the demand for manufactures between 1815 and the 1840's did not expand so rapidly, in relation to total factor-supplies, in Britain as in America. But there is no doubt that the expansion of the market was sufficient to warrant very large absolute increases in English industrial capacity. Quite apart from any increase in the efficiency of any sector, a considerable increase in income was achieved simply by shifts from less productive to more productive sectors. Even an industry without technical progress could therefore expect an increase in demand for

[1] *Industrialisation and Foreign Trade* (League of Nations, 1945), p. 13.

its products. Over and above any increase in total income; there was an increase in the demand for the products of factory industry at the expense of English domestic industry. But English factory industry also expanded at the expense of the domestic systems of other countries, that is foreign demand for English goods increased even in the absence of an increase in foreign incomes.

By the 1840's it was becoming increasingly difficult to expand exports further by these means; the net barter terms of the trade were turning against the United Kingdom. Hence the demand for free trade as a measure to increase the income of the foreign customers of England. But at the end of the 1840's—just when the possibilities of expanding exports at the expense of continental domestic industry were dwindling—a major boom started in several parts of the world associated with the gold discoveries, wars, and above all, railway building. This meant a very rapid increase of incomes in the countries to which Britain exported and, since Britain had very marked advantages as a producer of the consumer and capital goods they required, there was a spectacular expansion of British exports right down to the early '70's. The 1850's and '60's were a period in which incomes abroad were increasing much more rapidly than in the first half of the century, and when the marginal propensity of foreign countries to import from Britain was still very high; indeed, in the case of countries with heavy programmes of railway building, it may have been rising. The fears about British exports which had begun to be entertained in the '40's disappeared, and did not revive again until the later 1870's and '80's.

Thus for the first six or seven decades of the nineteenth century market prospects in England were favourable to the creation of new capacity; there were vigorous sources of demand for U.K. goods. This increased demand was not, of course, independent of the technical efficiency of English industry, but given the initial English superiority it proceeded even in the absence of technical progress, and it warranted an increase in capacity which gave English manufacturers excellent opportunities of adopting new methods and gaining experience.

After the 1870's the market prospects for British goods were less favourable to the creation of new capacity. In the first place, once England had become an industrial state, the possibilities of an increase in income from intersectoral shifts within the English economy were much more limited. In the second place, so far as foreign markets were concerned, though the increase in their real income was considerable, particularly in the U.S.A. and Germany, there was a decline in their marginal propensity to import from the U.K. The propensity would in any case have declined as these countries developed their own manufacturing industry, but the speed of the decline was accelerated by the tariffs which these countries imposed, and also by the fortuitous incidence of technical discovery in the key industry of the period—steel.

It is true that in certain regions of recent settlement the marginal propensity to import British goods remained very high. In some of these areas market goodwill probably retained for British goods a stronger position than was warranted by the price of British manufactures compared with those of other industrial exporters, for example in Australasia, S. America and India. But the rate of increase in income in the primary producing regions between the 1870's and the '90's was probably less than it had been in the '50's and '60's and certainly less rapid than it was in those economies which were industrialising.[1] Moreover, in S. America the preference for British goods was weaker than in most other primary-producing regions; in the 1880's when both Australia and the Argentine had development booms, the latter drew a smaller proportion of her additional imports from Britain. The Canadian preference for British goods was also weaker, and during the boom in the first decade of the twentieth century, a large part of her increased imports came from the U.S.A. These development booms in the Argentine and in Canada redounded to the advantage of British exports less directly than the early development booms in primary-producing regions. Though, directly and indirectly, they increased the demand for British exports, a significant part of the income which they created was probably employed (particularly by Americans) in repaying former borrowings from Britain, and from the point of view of British exports was dissipated. In the closing decades of the century, therefore, English industry as a whole was less assured of an increase in demand which would warrant the creation of a large amount of new capacity.

[1] A. J. Brown, *Industrialisation and Trade* (1943), pp. 60-61.

In Germany and the U.S.A. market prospects were much more favourable in the later decades of the nineteenth century than they were in England. In both there were large reserves of unexploited natural resources which had been developed rapidly as soon as the areas were supplied with railways. In Germany, too, simply because her industrialisation was late, the possibilities of increasing income by intersectoral shifts were still considerable. And the tariffs, which curtailed British exports of certain goods, concentrated the increase in domestic demand in Germany and America on local production. The fact that their total capacity was growing more rapidly meant that these countries had more opportunity of trying out new ideas, at a time when there were many to be tried out.

The argument applies most clearly to iron and steel where an inevitable long-term decline in foreign demand for British bulk products was greatly accelerated by technical and geographical discoveries and by foreign tariffs. The Gilchrist–Thomas process sharply altered the comparative advantages against Britain as a manufacturer of steel. It was more suited to German than to British ores because Britain's phosphoric ores (especially those of Cleveland) proved not to have *enough* phosphorous. Even if British ores had been just as suitable, the Gilchrist–Thomas process would still have produced a shift in comparative advantage; Britain's advantages for the production of acid steel had been very great, since it could import non-phosphoric ores so much more easily than Germany. Up to the Thomas–Gilchrist discovery Britain could produce acid steel and Germany could not; after the discovery Britain was, at best, no worse off as a steel producer and Germany was a great deal better off. The discovery of the Mesabi ores in the U.S.A. had a comparable effect; however enterprising English steel masters had been, there was bound to be a decline in the American demand for English iron and steel products.

This decline was very greatly accelerated by the imposition of tariffs in the 1870's and '80's. No doubt England's exports of the bulk iron and steel products would, in any case, have sooner or later declined, simply because the market for rails was declining, and other countries were becoming more efficient rail producers. The point about the tariffs was that they greatly accelerated the process. The British iron and steel industry had been heavily dependent upon exports—in the boom of the early '70's about half the British production of pig-iron was exported, either as pig-iron or as semi-manufactures—and particularly on exports to the U.S.A. and to Europe. The tariffs reduced two large markets for crude iron and steel products (especially rails). They also accentuated the cyclical fluctuations in the British industry. When, despite this tariff, the U.S.A. had to import steel rails—as in the boom of the early 1880's—the American industry made very high profits. Once the U.S. market fell off for cyclical reasons, British steel was virtually excluded from the American market. Conversely, the tariffs ensured that the expansion of American and German demand mainly benefited their own industries, and that they also enjoyed a steadier market than the British.

The net effect of these changes was to make new investment in iron and steel less profitable in Britain than in Germany and America during a period when many new technical methods were available to be incorporated. The effect is most clearly seen in the case of pig-iron production where the capacity in existence in Britain at the end of the 1870's was almost large enough to produce the output which was called for over the next twenty years; the output of 1882–3 was not surpassed until 1895. The British iron industry therefore had few opportunities of introducing new methods and the steel industry had a technologically backward base. Furthermore many of the improvements in steel were associated with improvements to the blast furnace, siting it in relation to the converters, making use of waste gases; and the fact that market prospects warranted the building of few new blast furnaces curbed the adoption of new methods in steel, at least in the Bessemer section which concentrated on rails. In the U.S.A. and Germany market prospects warranted a high rate of investment which facilitated the invention and adoption of new techniques. 'The up-to-date character of many American (steel) works is as much an effect as a cause of the expansion of the industry in America'.[1]

Market prospects are a less plausible explanation of the

[1] S. J. Chapman, *Foreign Competition*, *op. cit.* p. 92. There was undoubtedly some investment in the English industry, which in retrospect, can be seen to have been ill-advised: the heavy investment in malleable iron, in the boom of the early 1870's which made it more difficult later for steel to make headway; the over-concentration on railways, especially in the Bessemer section. But these mistakes were not the result of apathy or lack of technical skills. I am indebted for ideas on this industry to discussion with Mr J. F. Wright.

British lag in the new industries. But in the most important of these industries—electricity—differences in demand do seem to explain a good deal of the difference between the British and American experience.

In the first place, at the time when lighting by electricity first became a practicable proposition, there were more people in the U.S.A. than in Britain who were installing lighting for the first time, and whose choice between gas and electricity was therefore based upon a comparison of the total costs of the two methods. This was partly because, on the eve of the introduction of electric lighting, England was better lit than America; the English gas industry was much more efficient than its American counterpart and, before the coming of electricity, had tapped a larger part of the existing market for light. Furthermore, the demand for lighting was principally an urban demand, and the urban population of America in the later decades of the nineteenth century was growing more rapidly than that of Britain. For these reasons the total demand for lighting grew more rapidly in the U.S.A. than in Britain.

In the second place, in competing for the market in lighting, electricity was in a stronger position *vis-à-vis* gas in America than in Britain, because of the high price of American gas in relation to electricity, due principally to the high cost of American coal, the greater degree of monopoly in the American than in the English gas industry and the technological backwardness of the American industry. Nor was electricity the only case in which a new product and process had to make headway in England against competition from close substitutes produced by older industries or processes.[1]

There is another circumstance, more directly related to Britain's early start, which made it difficult for British manufacturers to incorporate new techniques. Much of the industrial capacity of England in, say, the early 1870's, had been built up gradually and there was therefore no particular reason why it should be well adapted for the absorption of new techniques. Many improvements available in steel in the 1880's and '90's depended on the concentration on a single site and in the same ownership of coking ovens, blast furnaces and rolling mills; but much of the English industry had been

[1] I am greatly indebted for enlightenment about electricity to Mr Ian Byatt who is writing a history of the industry before 1914.

created before the need for such concentration was evident. Similarly the introduction of the automatic loom in the cotton-textile industry involved considerable changes in the structure of the industry. The point has been well put by Mr Marvin Frankel. 'Real capital, especially in the later stages of industrialisation, may be made up of a number of components such that they cannot be replaced separately, but only on an all or nothing basis'.[1]

It is probable also that in certain industries the nature of the market for English manufactures—as distinct from its rate of growth—was unfavourable to the introduction of new techniques, since the purchasers of English goods demanded less homogeneous and standardised products than did American customers. The orders for English rails, for example, were mostly small and they involved many different sizes. This meant that, in some cases, the individual concern's output of a particular line of product fell below the minimum necessary for the economic installation of the latest techniques. Thus the fact that the English steel manufacturer needed a larger stock of rolls and had to make frequent changes was held to preclude the adoption of such innovation as three-high mills.[2] Moreover, in the iron and steel industry, because demand was unhomogeneous, each concern tended to have its separate goodwill market, and this made it easier for high-cost producers to resist elimination. We have already considered some of the effects of imperfection in the product market in the case of the U.S.A. in the early nineteenth century. But the effects are very greatly influenced by the rate of growth of the total market. Because of the slow growth of English industrial capacity in the last three decades of the nineteenth century, the adoption of new techniques was much more dependent than in America on re-equipment by existing firms and the expansion of the more—at the expense of the less—efficient, and it was precisely this process which was impeded by market imperfections.

The unhomogeneity of demand for British goods, though mainly due to the geographical diversity of our markets, was

[1] M. Frankel, 'Obsolescence and Technological Change in a Maturing Economy', *American Economic Review*, June 1955, XLV, pp. 296–319, together with the *Comment* by D. F. Gordon, and Frankel's reply in *Am. Ec. Rev.* September 1956, XLVI, pp. 646—56.

[2] D. L. Burn, *op. cit.* p. 192.

also related to the rate of growth of demand. It was pointed out in the discussion of American development, that new demand is more readily standardised than demand diverted from other suppliers or products. In England a substantial part of demand was of the second type.[1] Thus, in the first half of the century, factory production of cotton textiles increased partly by invading markets already served by industry organised on a putting-out system, that is by serving customers whose tastes were already formed. In the second half of the century, a good deal of the steel output went to replace malleable iron, so that many of the imperfections in the market for iron products were perpetuated in the market for steel products. But there is the further point that it was easier for manufacturers (or merchants) to impose uniform standards on customers in a sellers' market, when demand was increasing rapidly in relation to capacity. In the English cotton-textile industry, capacity usually responded very promptly to demand, since the units of operation and ownership were small and entry into the industry was relatively easy. The customers of a Massachusetts cotton-textile industry in the early nineteenth century had sometimes to take what they could get; the English industry almost always faced a buyers' market, and manufacturers and merchants were more often anxious to attract demand by meeting consumers' tastes than they were able to impose standard types of product upon the market. This was particularly the case in the last quarter of the century when the output of the industry grew more slowly than it had done previously. Similarly if, in the later nineteenth century, there had been a very rapidly growing new demand for English steel products, English producers would have more easily been able to standardise their products, and by producing standard products at lower prices they might have been able to condition the old demand.

[1] This situation, of course, occurred in all European countries with a tradition of mechanical crafts. Emil Rathenau, in an autobiographical sketch on his visit to the Philadelphia Exhibition of 1876, commented on the lack of standardisation of German machines compared with the American, and ascribed it to the character of demand in Germany. The German purchaser of machines was usually better informed than the seller about his precise requirements, and he made his order dependent on the carrying-out of his demands. He sought, by paying extra, to obtain machines specially adapted to his circumstances and this necessitated very careful construction of individual machines. It would be better, Rathenau thought, to come to an agreement about a standard type of machine. A. Riedler, *Emil Rathenau* (Berlin, 1916), pp. 28–9.

Some of the reasons we have given for England's smaller increments of new investment in the later nineteenth century, compared with America's, arise from England's early and long-sustained start. It is not argued that England was on balance a loser from her early start. Because of this start England's income in the 1870's was higher than it would otherwise have been (and higher than that of other countries except the U.S.A.) and her savings were much larger. Because she had already a large industrial plant she was able to use her savings to build more and better houses and to provide the regions of overseas settlement with their complement of public utilities. But because she already had a large industrial plant which in many branches was adequate to the demands made upon it in the '80's and '90's, the incentive to install new capacity and the opportunity of trying out new methods were circumscribed. And in explaining the different rates of technical progress this fact is of crucial importance.

The same circumstances may also account in large part for the quality of British entrepreneurial performance in the same period. The entrepreneurial deficiencies to which the performance of the British economy is attributed can plausibly be explained as a consequence of that performance. The slow rate of expansion of British industry affected the performance of business men. Great generals are not made in time of peace; great entrepreneurs are not made in non-expanding industries. Because the market was growing slowly, the risks of adventure were much greater than in Germany and America, as may clearly be seen from the fate of those who showed enterprise in the electrical industry in the early '80's, and came croppers as a result.

But the slow expansion must also have affected the recruitment of business men. An Englishman's choice of career was, it is true, very much influenced by tradition, convention and inertia, and no doubt in England these tended to channel talent away from business towards the professions. But it was also responsive to changes in reward,[1] and after the 1870's several purely economic circumstances combined to divert men of ability away from business. The high profits of the 1850's and

[1] That the supply of recruits to professions and trades as a whole was responsive to changes in reward was assumed by the opponents of differentiation in the income tax. F. Shehab, *Progressive Taxation* (Oxford, 1954), pp. 95–6.

'60's had attracted an unduly large number of entrants into business so that even had the rate of profit been sustained in the following decades, the rewards of individuals would have fallen. But, over and above this there was a fall in profits, a decline in the ability of firms to offer attractive salaries; in periods of falling prices business was at a disadvantage *vis-à-vis* the professions in competing for talent. The overflooding of an occupation which in any case was becoming less popular meant that business was very much less attractive for men who were making their choice of profession from the 1870's on. Where there was lack of growth there were few new firms, the average age of men at the top of existing firms was high and nepotism was most likely to occur.

The more able men there are in an industry the more chance the industry will throw up an entrepreneur of genius. A country where several industries throw up a genius, gives the impression of being endowed by nature or by its social order with an exceptionally large amount of entrepreneurial talent. And this was the impression which the U.S.A. gave. But the abundance of entrepreneurial talent in the U.S.A. was the consequence rather than the cause of a high rate of growth; and it was the slow expansion of English industry which accounted for the performance of English entrepreneurs in the later nineteenth century not vice versa. Where market conditions were favourable to the expansion of capacity British business men were just as venturesome and dynamic as the American. Charles Parsons in marine engineering, Alfred Jones, the master-spirit of Elder Dempster, who acquired control of the West African carrying trade and brought the banana to the British breakfast table, William Lever, who exploited the mass market in soap, Harmsworth, who exploited the coincidence of literacy and technical changes in printing: these men are the equal in enterprise and achievement of any American entrepreneurs. They were all operating in fields where circumstances quite independent of their own initiative were favourable to expansion. In the provision of merchant shipping and financial services there was considerable entrepreneurial achievement, to judge from the expansion of the U.K.'s income from invisible exports. But perhaps the best instance of the relation of investment to entrepreneurial ability comes from the steel industry. For while the Bessemer section was backward, the open-hearth section, which was expanding because it serviced the shipbuilding industry, was technologically vigorous.[1]

There were, of course, differences in entrepreneurial ability and performance which cannot be explained in these terms. At the start of the electrical industry the leading names were American and continental European. There was no one in England to equal Brush and Edison. This was an entrepreneurial and not primarily a technical deficiency; it was not so much that the Americans were better engineers as that they had greater insight into the nature of the market for electric light, and employed their technical skill in perfecting a system which met the requirements of the market. It is possible also that a difference in pure entrepreneurial ability is reflected in the comparative history of the motor-car in the two countries. In the cotton-textile industry, as an English expert observed in 1902, 'when every allowance has been made for the differences in their situations the conclusion is hard to resist that the average English manufacturer is more cautious—more timid if you will and more liable to reject a good thing because it happens to be new and hitherto untried, than the average American manufacturer'.[2] The slow adoption of the automatic loom in England cannot be explained entirely by factor-proportions, for it was made and worked, not only in America, but in Germany, Austria, France and Belgium.

But was the difference of entrepreneurial ability all that great? In electricity, the men who developed the industry in England, Swan, Crompton and Siemens, were all men of considerable ability and very little behind the Americans technically. The difference in the development of the internal combustion engine was greater: there was more experiment in America than in Britain. But it may be doubted whether this kind of disparity in 'pure entrepreneurial ability' mattered a great deal in the later nineteenth and early twentieth centuries. Where there were no other impediments, it did not prevent the rapid adoption of methods pioneered elsewhere. The absence of the higher forms of inventive and entrepreneurial ability represented by Brush and Edison in the electrical industry was not a major handicap to the development of the

[1] W. A. Sinclair, 'The Growth of the British Steel Industry in the Late Nineteenth Century', *Scottish Journal of Political Economy*, VI, No. 1 (February 1959).

[2] Young, *op. cit.* p. 137, XV.

British industry, since the Brush and Edison systems were very rapidly imported. Where advanced technology of foreign origin was not incorporated, the explanation is probably more often the limitations of economic circumstances rather than the inertness of British entrepreneurs.

If, as we have argued, technical progress in England was retarded by the slow growth of demand for British exports in the last three decades of the century which afforded relatively few opportunities for the growth of new capacity, why, it may be asked, did the boom in exports after 1900 not stimulate technical development? In certain industries the answer is probably that the British lag was, by then, too substantial to be made up easily, because of the lack of the know-how which would have been accumulated by keeping more continuously in touch with new methods. Partly also the explanation may be that the increase in export demand was principally for those goods where the possibilities of technical progress were slightest. It was in the new industries that the possibilities were greatest. Their methods were more elaborate than those of the older industries, and the potentialities of these methods had been less fully exploited. They were the industries with the best long-term demand prospects, and the least stereotyped labour-force.

We shall now attempt to sum up the influences on the technological history of the U.S.A. and Britain in the later nineteenth century, and particularly those most directly responsible for the contrasts between them.

There was no significant difference in the stock of fundamental ideas available to technologists. In this period, Britain and the U.S.A., along with Germany and France and some of the smaller European countries, shared a common stock of such ideas in the sense that there were many people in each country who were intellectual masters of the notions underlying the new developments in technology. From this point of view, these countries form a common group clearly distinguished from most of the rest of the world.

There was certainly some difference in the stimulus to derive new processes from the common stock of fundamental ideas. The rapid growth of the American economy provided incentives to search for new methods in all directions and dear American labour an incentive to search especially for labour-saving improvements. Possibly too American entrepreneurs were more responsive to a given stimulus for non-economic reasons, for example because society at large attached greater prestige to technological achievement. It may be, that is, that where the Americans drew more readily upon the common stock of ideas it was because they were more anxious than the British to adopt new methods. But much more important than differences in the desire to adopt new methods were differences in the ability to do so, and the essence of the argument of this chapter has been that it was easier to adopt new methods in America than in England.

There is not much force in the argument that the English entrepreneur in this period was hampered by the absence of the scientific skills required by the new technological developments. He was no worse off (indeed he may have been better off) than his American counterpart, and even in comparison with the Germans his disadvantages on this score are commonly exaggerated. It is often said that a shortage of chemists—the result of deficiencies in the English educational system—was responsible for the failure of the English dye-stuffs industry, after a promising early start, to make progress in the 1880's and 90's. No doubt there is something in this. But even in this case, a major part of the explanation is that the English industry failed to attract or retain the available scientific ability, and lacked the desire to train its own scientists, because its prospects deteriorated for reasons independent of the supply of scientific skills. The English patent law did not compel a patentee to work his patent, and this enabled the Germans to deny to English manufacturers the use of the most important German inventions in this field. The English, being thus restricted to a narrow range of dyes, were more vulnerable to change and less likely to make new discoveries than their German counterparts; and under a free-trade regime, once German manufacturers had established even a slight margin of superiority, there was no reason why English customers should not buy their dyes from Germany.[1] Thus a lead once lost was extremely difficult to recover. In other industries I do not believe that the ability of the English entrepreneur to innovate, in comparison with that of his rivals,

[1] L. F. Haber, *The Chemical Industry during the Nineteenth Century* (Oxford, 1958), pp. 198–204.

was seriously limited by the terms on which he could obtain the appropriate scientific and technical expertise.

Where there were serious limitations on innovation they arose principally from the market facing the English entrepreneur. There is first of all the question of the size of the market. It has often been argued that the extent of the American domestic market, unimpeded by tariff barriers, made methods profitable in the U.S.A. which would not have been profitable elsewhere: the greater the total volume of production in a country, the greater the specialisation of products and the greater the inducement to mechanise their production.[1] But while it may be true that the larger economy can more easily avail itself of the advantages to be derived from specialisation and mass production, there is no reason to believe that the differences between Britain and the U.S.A. at the end of the nineteenth century were significant from this point of view. The important difference was not in the size of the market but in its rate of growth.

For reasons we have already explored, American industrial capacity grew more rapidly than the English, and thus American industrialists had more opportunities of trying out new methods. And this consideration does not apply only to America. All the countries who followed the English into industrialisation in the later nineteenth century grew more rapidly than the English because there was a large back-log of technology for them to absorb and a great volume of resources to be shifted into modern industry from agriculture and domestic industry. In the case of some of the late industrialisers, for example Japan, most of their effort was absorbed in copying existing technology; and where technical and scientific skills were very scarce it was probably more economic to devote them to imitation and adaptation rather than to technical innovation. But the rapid growth of capacity among the late-comers, even where it was principally imitative, afforded opportunities for testing new methods; and in the U.S.A. and Germany there were sufficient supplies of technical skill to allow them to make indigenous improvement *pari-passu* with the absorption of existing techniques.

This was the first serious limitation on the comparative

[1] A. Young 'Increasing Returns and Economic Progress', *Economic Journal*, XXXVIII (1928), 540.

ability of British entrepreneurs: the slower rate of growth of the economy. The slower rate of growth was the more important because the structure of industry imposed more severe restraints on the choice of techniques in Britain than in the U.S.A. British industry was longer established and the entrepreneur more limited by history and by the need to shape his investments 'to fit the inherited structure of complementary assets'.[1] Difficulties arose on this score even when a particular improvement was within the power of the individual entrepreneur to adopt. For example, the size and layout of an old textile mill imposed restrictions on the choice of techniques which were not present in one newly built in knowledge of the most recent techniques. But many improvements were profitable for the individual concern only if complementary changes were made in other concerns; for example, an increase in the size of ships might be worth undertaking only if docks were enlarged. This type of change was particularly difficult to effect in later nineteenth-century Britain because of the complex division of labour between processes and industries which had developed during the long period of British industrial supremacy. Division of labour, it has often been observed, promotes mechanisation because the more specialised, and hence simpler, processes which result lend themselves more easily to the use of machinery than a group of complicated, because undifferentiated, processes. But there is another side to the coin. The fact that division of labour had been carried so far in English industry, and that the structure of inherited assets was so complicated, reduced the number of improvements which were within the ability of a single concern to introduce.

There are other reasons why such complementary improvements were easier to bring about in the U.S.A. For one thing American industries were under a common pressure to standardise their production and shared traditions of standardisation. Then again the strength of the merger movement in America was favourable to adoption of improvements which cut across firms. Mergers and amalgamations are ways in which an entrepreneur can assure that such improvements are made; and, because the merger movement in the later nineteenth century was weaker in Britain than in America, the opportunities for securing

[1] G. B. Richardson, *Information and Investment* (Oxford, 1960), p. 114.

complementary improvements were more limited. For this weakness of the merger movement there are, of course, many reasons; but one was certainly that mergers were more difficult to arrange in long-established industries and between firms with a complicated and deeply-entrenched specialisation of market and product.

Besides such economic reasons for differences in technical progress, there were no doubt social and psychological reasons. A society in which existing habits and institutions were widely accepted and freedom slowly broadened down from precedent to precedent found it easier to maintain political and social stability; but it was also likely to grow more slowly than a society where individuals were bent on rising in the world and were ready to break with old ties and customs in order to do so. The restlessness of American life was favourable to economic achievement.

There is another consideration of a similar kind. Societies are more likely to grow rapidly, it is often argued, if they feel they require growth for more than purely economic reasons, that is if they are moved by an ideology which gives their leaders a strong emotional commitment to the task and shakes up the feelings and routines of the population as a whole. Various beliefs have served this purpose at different times. Thus it has been suggested that Calvinism provided an ideological reinforcement to economic effort in the sixteenth century. And in the nineteenth century, among countries that lagged behind Britain in industrialisation, the tension between their existing state and their potentialities—the desire to catch up—generated ideologies favourable to development, such as Saint-Simonism in France, or turned existing currents of feeling to this end, for example nationalism in Germany. In our own day, the need to restore economies shattered by war has fired Europeans with a fervour for the task—a fervour which, together with the disruption of old routines by the war, goes some way to explain why, since 1945, France and Germany have grown more rapidly than countries like Britain and the U.S.A. which have not had to meet the challenge of rebuilding large parts of their industrial capacity from scratch.

In the period in which we are particularly interested in this chapter, ideological influences were more favourable to growth in the U.S.A. than in Britain. In the U.S.A. the need to open up the vast natural resources of the country had roused general feelings of buoyancy and optimism. In addition immigration provided an emotional dynamism. For the Europeans who had taken the momentous step of tearing up their roots and migrating to America had an immensely strong incentive when they got there to prove their decision right by ensuring, through hard work, that they did in fact better themselves. In Britain, on the other hand, the mid-Victorian pride in her industrial leadership —a pride which had its basis in fact, but which developed a life of its own and acquired an ideological character—was blighted by doubts in the depressions of the 1870's and '80's; and such ideological movements as there were in Britain in the decades before 1914—imperialism and socialism for example—were not particularly favourable to economic effort.

I know of no way of forming more than a general impression of the relative importance of these non-economic influences. There are some contrasts which are best interpreted in such terms. But I do not believe that this is true of Britain and the U.S.A. in the later nineteenth century. The differences in their technical progress, though significant enough to require explanation, were relatively small if we take a broad view of economic history; and there were, of course, fields in which Britain was ahead. Such lags as there were in the adoption of new methods in British industry can be adequately explained by economic circumstances, by the complexity of her industrial structure and the slow growth of her output, and ultimately by her early and long-sustained start as an industrial power.

INDEX